Individualized Instruction in Mathematics

compiled by

SAM DUKER

The Scarecrow Press, Inc.

Metuchen, N.J.   1972

Other Scarecrow Press books by Sam Duker

Listening: Readings. 1966.
Listening Bibliography. 1964. (2nd ed. 1968.)
Individualized Reading: An Annotated Bibliography. 1968.
Individualized Reading: Readings. 1969.
Listening: Readings. Vol. 2. 1971.
Teaching Listening in the Elementary School: Readings. 1971.

Library of Congress Cataloging in Publication Data

Duker, Sam, 1905-   comp.
    Individualized instruction in mathematics.

    Bibliography: p.
    1. Mathematics--Study and teaching--Addresses, essays,
lectures. I. Title.
QA11.D79   372.7   72-5739
ISBN 0-8108-0533-2

Copyright 1972 by Sam Duker

DEDICATED

with deepest affection

to Corinne

without whose help, given when it was most needed,
this book and innumerable other things might never
have been accomplished.

TABLE OF CONTENTS

PREFACE AND ACKNOWLEDGMENTS

The material which appears in this book is intended to present the reader with thought-provoking descriptions of ways in which the teaching of mathematics may be individualized. In addition, the articles included present a certain amount of discussion and explanation of the theory on which individualization of any sort of instruction is based. The examples given of individualized teaching of mathematics at various school levels have not been selected on the basis of successful implementation of the ideas presented. Many of the examples contain data showing that the experiment or method was not in that particular instance as effective as more conventional methods. This is done since this book does not presume to be a guide in the sense of a teaching manual. It is rather intended to present information which will allow each reader to come to his own conclusions.

Since my own interest in individualization of instruction stems largely from the two years I spent as a teacher of a one-room rural school a quarter of a century ago, I must first of all acknowledge my deep sense of gratitude and appreciation to the children who came to that little school to learn, to the parents who sent their children there, and to the community which supported the school. Obviously in such a school where one finds from 15 to 25 children covering the range from the first to the eighth grade, individualization becomes a necessity regardless of the teacher's desires or training. Most of my colleagues at neighboring rural schools of the type in which I taught had had only a very minimum of teacher training, but the fact that they had to individualize their instruction was taken for granted and few of them seemed to find this facet of their teaching much of a problem. Nowadays when I hear highly trained teachers saying that they like the theory but don't think that it is possible to implement it in a class of thirty (all in the same grade, incidentally) I am tempted to believe that they should have had a little experience in the rural school of yesterday.

8

Those two years I spent (on the plains of Oklahoma) have left a lasting impression with me and have colored my educational philosophy to an extent I cannot adequately describe. These years were the greatest in my life as far as learning is concerned. In compiling a book like this I therefore find acknowledgments an inadequate way to describe the influence of the people of that poverty-ridden, land-eroded school district. I will not say that the patrons of the school always either understood or approved what I was doing but there was very little unkind criticism but a great deal of encouraging interest, and in some instances even outright approval. For all this I am humbly grateful.

Since leaving that country school, it has been my privilege to teach at colleges where my colleagues on the faculty were in the forefront in seeking ways to improve teaching. While not all of them would agree that individualized instruction is the most desirable answer to providing for individual differences, all of them shared with me a deep concern about accommodating such differences in one way or another. My acknowledgment of the part they have played in my thinking (which is reflected in a book such as this) is therefore not merely pro forma but comes from a sense of grateful appreciation for the many ideas I have received from discussions with them.

A very distinct influence on my thinking about individualization was exerted on me by Dr. May Lazar, now retired. During her years spent in the Bureau of Research of the New York City Board of Education, she had a tremendous influence on the actual implementation of programs on the individualized teaching of reading in literally thousands of classes in the schools of New York City. I am profoundly grateful to this valued friend for inspiration and learning about the process of individualized instruction.

The formal acknowledgments made to the authors of the material I selected for inclusion in this collection seem to me to be a very inadequate way of expressing my appreciation to them. Not only did I learn much from reading their material, but I was also tremendously impressed with the gracious way in which the authors, without exception, consented to my using their material. I also appreciate their patience and forbearance during the long period of delay between the time they gave their consent and the final publication of this book.

In order to compile the material for this book it was

necessary to read many items. Most were not finally included; this does not mean that the quality of the material was
bad but only that it either duplicated what I already had
selected, or for one reason or another did not fit into my
plans for this particular anthology. I learned much from
this reading and this book is richer for my having read. Obviously it is not practicable to name the authors in question
but my acknowledgment to them is no less sincere because
of that.

The planning of this work took place largely in the
Brooklyn College Library. I am deeply obliged and indebted
to the staff members of that library for much help willingly,
expertly, and efficiently given. There are many others who
have helped in one way or another. Their names are legion
but I assure each of them that I do remember and appreciate
their efforts.

Specific Acknowledgments

Professor Elaine Bartel of the University of Wisconsin kindly gave me her permission to use excerpts from her
1965 University of Wisconsin doctoral dissertation entitled
"A Study of the Feasibility of an Individualized Instructional
Program in Elementary School Mathematics."

Excerpts from a 1968 University of Michigan doctoral
thesis entitled "Provisions for Individual Differences in Seventh Grade Mathematics Based on Grouping and Behavioral
Objectives: An Exploratory Study" are used with the kind
consent of the author, Dr. James E. Bierden.

Robert E. Botts of Long Beach, California, and the
editor of the Journal of Secondary Education gave me their
permission to use excerpts from "The Climate for Individualized Instruction in the Classroom" which appeared in November 1969.

Professor Richard M. Bradley was kind enough to
allow me to include in this collection portions of his 1967
Temple University thesis which he wrote for his doctorate,
"An Experimental Study of Individualized Versus Blanket-
Type Homework Assignments in Elementary School Mathe-
matics."

Dr. Donald Deep gave me permission to include in

this book excerpts from his doctoral thesis completed at the University of Pittsburgh in 1967 and entitled "The Effect of an Individually Prescribed Instruction Program in Arithmetic on Pupils at Different Ability Levels."

Professor Paul Douglass of Rollins College in Florida granted me permission to include in this book portions of his article, "Theory, Practice, and Perils of Independent Education"; consent to its use was also given by Dr. Demer M. Goode, editor of Improving College and University Teaching, in which this article appeared in 1968.

Dr. William T. Ebeid completed his doctoral dissertation, "An Experimental Study of the Scheduled Classroom Use of Student Self-Selected Materials in Teaching Junior High School Mathematics," at the University of Michigan in 1964. He kindly sent me his permission to use portions of his work in this book from Cairo, where he is presently professor of mathematics education at Assuit University.

The article, "Toward Individualized Learning," by Barbara Bree Fischer and Louis Fischer, which appeared in the Elementary School Journal, is included with the kind permission of the senior author and of the University of Chicago Press, publishers of the Elementary School Journal.

Dr. Jack R. Fisher has generously allowed me to use selected portions of his 1968 University of Pittsburgh doctoral thesis, "An Investigation of Three Approaches to the Teaching of Mathematics in the Elementary School."

Excerpts from a 1966 Indiana University dissertation, "The Relative Merits of Selected Aspects of Individualized Instruction in an Elementary School Mathematics Program," are used with the permission of the author, Dr. Victor L. Fisher, Jr.

Professor Frances Flournoy of the University of Texas has kindly permitted me to include portions of her article, "Meeting Individual Differences in Arithmetic," which appeared in the February 1950 issue of the Arithmetic Teacher.

The Arithmetic Teacher article, "Through the Years: Individualizing Instruction in Mathematics," is reproduced in this collection with the kind permission of the author, Professor E. Glenadine Gibb of the University of Texas, and of Dr. Carol V. McCamman, managing editor of the National Council of Teachers of Mathematics, publishers of the <u>Arithmetic Teacher.</u>

Excerpts from the article by Professor Charles J. Gorman of the University of Pittsburgh entitled "The University of Pittsburgh Model of Teacher Training for the Individualization of Instruction," which appeared in the Spring 1969 issue of the <u>Journal of Research and Development in Education,</u> are used with the consent of both the author and Mrs. Evelyn J. Blewett, executive editor of the <u>Journal.</u>

Professor William A. Graham of Florence State University and Dr. Carol V. McCamman, executive editor of the National Council of Teachers of Mathematics, publishers of <u>Arithmetic Teacher,</u> both graciously consented to the use of excerpts of an article, "Individualized Teaching of Fifth and Sixth Grade Arithmetic," which appeared in the April 1964 issue of <u>Arithmetic Teacher.</u>

Dr. Jettye Fern Grant has kindly allowed the use of excerpts from her 1964 University of California dissertation, "A Longitudinal Program of Individualized Instruction in Grades 4, 5, and 6."

The article, "An Analysis of Individual Differences in Arithmetic," which appeared in the November 1964 issue of <u>Arithmetic Teacher,</u> is excerpted with the kind permission of the author, Professor Oscar T. Jarvis of the University of Texas at El Paso, and of the publisher, the National Council of Teachers of Mathematics.

Selected portions of an article by Marilyn Jasik, which appeared in <u>Childhood Education</u> of October 1968, entitled "Breaking Barriers by Individualizing," are used with the permission of the author and of the publisher, the Association for Childhood Education International.

Selected portions of "Individual Differences," which appeared in the <u>25th Yearbook of the National Council of Teachers of Mathematics in 1960,</u> are reproduced here with the kind permission of the authors, Professors R. Stewart Jones and Robert E. Pingry, both of the University of Illinois, and of the publisher, the National Council of Teachers of Mathematics.

"Achieving Individualized Instruction," which appeared

in the Winter 1969 issue of High Points, is included in this
anthology with the approval of the author, Abraham Kaplan,
and the publisher, the Office of Educational Publications of
the Board of Education of the City of New York.

Professor Eugene R. Keffer of Central Missouri
State College has given his permission for the use of excerpts from "Individualized Arithmetic Instruction," an
article which he wrote for the May 1961 issue of Arithmetic Teacher.

Excerpts from The Borel Individualized System of
Instruction Program, a curriculum document written by
William R. Kramer, principal of Borel Middle School,
are included here by permission of the author.

Professor J. Murray Lee has kindly consented to the
use of portions of his 1954 Education article, "Individualized
Instruction," for which permission has also been given by
the Bobbs-Merrill Company, publishers of Education.

Professor Lucille Lindbergh of Queens College of
the City University of New York has graciously consented
to allow me to reproduce her 1964 article, "What is Individualized Education?' which appeared in Individualizing
Education, a pamphlet published by the Association for
Childhood Education International, which has also given permission for its use.

Selected passages from the doctoral thesis, "A descriptive Analysis and Evaluation of an Integrated Program
of Individualized Instruction in Cedar City High School,"
written at Brigham Young University in 1964, are used with
the kind permission of its author, Dr. J. Clair Morris.

Excerpts from an article by Maria Paul Morrison
which appeared in the December 1937 issue of Education
are used with the kind permission of the copyright owner,
the Bobbs-Merrill Company, publishers of Education.   I
have made diligent efforts to locate the author of this
superior article, "Mass Method Versus Individual Method
in Teaching Multiplication to Fourth Grade Pupils," but
have been unsuccessful; I have reluctantly included it without obtaining her permission, but I do wish to acknowledge
her part in this book.

Professor Cecil Nabors of Sam Houston University

has kindly agreed to allow me to include in this collection selected portions of his 1968 University of Houston doctors' dissertation entitled "The Effect of Individualized Verbal Problem Assignments on the Mathematical Achievement of Fifth Grade Students."

Excerpts from the doctoral dissertation completed by Professor K. A. Neufeld of the University of Alberta, at the University of Wisconsin in 1967, entitled "Differences in Personality Characteristics Between Groups Having High and Low Mathematics Achievement Gain under Individualized Instruction," are used with the kind permission of the author.

Dr. George C. Nix has kindly permitted the use of several portions of his 1969 Auburn University thesis entitled "An Experimental Study of Individualized Instruction in General Mathematics."

"A Study to Determine Whether Fifth Grade Children Can Learn Certain Selected Problem Solving Abilities Through Individualized Instruction" was written in 1966 by Dr. Raymond J. O'Toole as his doctoral thesis at Colorado State College. It has been excerpted for this collection with the permission of the author.

An article appearing in the April 1969 issue of the Journal of Secondary Education, published by the California Association of Secondary School Administrators, entitled "An Observation of Computer Assisted Instruction on Under-achieving, Culturally Deprived Students" is excerpted with the kind permission of its author, J. Marion Patterson, and of the publisher.

Caroline C. Potamkin has permitted the use of portions of her December 1963 article, "An Experiment in Individualized Arithmetic," as has the publisher, the University of Chicago Press.

Professor Helen Redbird of Oregon College of Education at Monmouth and the National Council of Teachers of Mathematics have kindly allowed the use of excerpts from "Individualizing Arithmetic Instruction," which appeared in the May 1964 issue of Arithmetic Teacher.

Excerpts from "New Mathematics Curriculum Called Part of 'Quiet Revolution' in Teaching Methods," which appeared in the April 1970 issue of Educational Development, published by the Regional Educational Laboratory for the Carolinas and Virginia, are used with the kind permission of the publisher.

Special acknowledgment is due Dr. Robert G. Scanlon, Program Director of the Individualized Learning Program of Research for Better Schools of Philadelphia. Not only has he allowed me to use excerpts from his 1967 University of Pittsburgh doctoral dissertation entitled "Factors Associated with a Program for Encouraging Self-Initiated Activities by Fifth and Sixth Grade Students in a Selected Elementary School Emphasizing Individualized Instruction," but he has also been very generous in providing me with materials and information that have been invaluable in the preparation of this book.

An editorial written by Professor Archibald Shaw, now of Michigan State University, which appeared in the September 1963 issue of American School and University, is reproduced with the permission of the author.

Professor Henry D. Snyder of the University of Wisconsin at Milwaukee kindly gave his permission to use excerpts from his doctoral thesis, "A Comparative Study of Two Self-Selection-Pacing Approaches to Individualized Instruction in Junior High School Mathematics," completed in 1966 at the University of Michigan.

Dr. Robert L. Spaulding of San Jose State College kindly allowed the use of selected passages from his article, "Personalized Education in Southside School," which appeared in the January 1970 issue of Elementary School Journal. The University of Chicago Press, the publisher, also gave permission for the use of this material.

An article by R. B. Thompson, "Advanced and Remedial Instruction in Mathematics," which appeared in School Science and Mathematics in 1941, is reproduced in this collection with the kind permission of the Central Association of Science and Mathematics Teachers, Inc., publishers of that journal. Although diligent efforts have been made to discover the present whereabouts of the author, I have been unable to reach him. In the absence of his permission, I nevertheless wish to express my grateful acknowledgment to him.

Walter L. Whitaker of the Culver City (California)

Unified School District wrote the article, "Individualized Arithmetic--An Idea to Improve the Traditional Arithmetic Program," which appeared in Arithmetic Teacher in 1962, and he has given me permission to use major portions of it in this collection.

Professor Bernard R. Wolff of Lewis and Clark College has graciously allowed me to use excerpts from his 1968 University of Oregon doctoral thesis, "An Analysis and Comparison of Individualized Instructional Practices in Arithmetic in Graded and Non-Graded Elementary Classrooms in Selected Oregon School Districts."

Professor John L. Yeager of the University of Pittsburgh permitted the inclusion in this collection of selected portions of his 1966 University of Pittsburgh thesis, "Measures of Learning Rates for Elementary School Students in Mathematics and Readings under a Program of Individually Prescribed Instruction."

Sam Duker

I. INTRODUCTION

This book is about the individualization of instruction in the teaching of mathematics. The emphasis of most of the content is on the elementary school level. I hope that upon completion of this book it will be crystal clear to the reader that the principles expounded here concerning the feasibility and desirability of individualized instruction are not confined either to mathematics or to the elementary school level. If in fact the theories and their implementation described in this book are sound, then they are practicable and desirable in the teaching of any subject matter and at any school level.

The material in this book represents a variety of viewpoints concerning individualization. I will not deny that I possess a bias in favor of this kind of approach to all instructional processes. It has been my sincere effort, which I hope has been a successful one at least to some degree, to select material which will constitute a fair and balanced presentation rather than merely articles that would reflect only my own biases.

After a lifetime, I have become convinced that the present, as compared to either the past or the future, is always a moment in time which presents the most acute crises and calls for the most urgent decisions. This is good, I think, because it is this sense of urgency and importance we attach to the present that is, as the proverb has told us, the mother of invention and a spur to the creativity and ingenuity called for in any attempt to reach viable solutions to the problems of the day.

It is unlikely that there are more than a very few of those connected in one capacity or another with the responsibilities of education who can be unaware of the sense of urgency evident in present-day attempts to find solutions to the problems facing the American school.

17

The last two decades have added substantially to the schools' problems in the 1970's by making them the testing ground for two gigantic social movements. Regardless of one's attitude toward these problems or one's opinion about their most desirable solutions, and certainly without reference to whether or not these problems should have become associated with the school, one matter is almost universally agreed on. Few would deny that court mandated desegregation and the court mandated separation of church and state have added to the problems of the American schools. Few doubt that both these matters will pose further problems for the school during the present decade.

One of the phenomena of the present day is the greater general participation by growing numbers of groups and individuals in the search for solutions to educational problems. Perhaps this is due to factors mentioned in the previous paragraph. Whatever its causes, in my lifetime I have never observed a sense of general concern greater than I see at present.

This is particularly true of students. From time immemorial, some degree of interest in what occurs in the process of education at the university level has been evidenced by students. At the University of Bologna, for example, many centuries ago there was a long period of time during which students completely dominated the governance of that institution. Obviously this experience was not fatal to the survival of that university since it has had the longest continuous existence of any contemporary institution of higher learning.

In the history of education we find innumerable instances of student participation of one sort or another in the governance of universities. The student of today, rebelling and protesting against the actions, policies, and decisions of the "establishment," whether it be that of the college or that of society in general, finds his counterpart all through the centuries during which universities have existed.

Today there appears to be an extraordinarily urgent demand by college students that they be given a part in the decision making processes at many levels of university life specifically including curriculum, faculty selection and retention, and budgetary planning. The pressure by students to obtain such rights has spilled over into the secondary school level. Here the demand is not so much to obtain the

right to participate in governance as it is to be allowed to
be freed of what they consider to be archaic and repressive
rules and regulations. The courts have become involved,
to a hitherto unprecedented degree, in settling disputes about
dress, personal appearance, and restrictions on free ex-
pression related to school publications, handbills, and the
like. The list of cases involving "hair" alone is several
pages long. Many high schools have had to close from time
to time for short or for longer periods because of student
unrest concerning actions by school authorities which were
regarded as irrelevant.

While I know of no instances of pupil motivated pro-
tests in the elementary school, anyone who thinks that life
has been peaceful and serene at that level is sadly mistaken.
During the past several years picket lines of indignant parents (often accompanied by their children) have become a
common sight as they effectively boycotted schools in many
parts of the United States. Transfers from school to school
with or without busing, as one example, have not been accepted as they were in the past and school officials have
been informed very firmly of this opposition.

All such activities could be regarded by the extremely
optimistic as a sign that the long-standing tradition of local
control of the neighborhood school is still a viable reality.
The difficulty is that such a view does not take account of
the ugly and violent nature of the criticism of the school today or of the deep-seated resentment that prevails in many
quarters concerning the policies and accomplishments of the
schools.

The militancy evident in school affairs today is by no
means confined to students and their parents. It extends
both to the general public, particularly the taxpaying sector,
and to professional staffs.

Only a few decades ago the stereotype of the teacher
was a meek, mousy type of spinster who made no protest
when asked to sign her teaching contract which provided not
only for scandalously low compensation but also for many
personal restrictions on her mode of life. Smoking, drink-
ing, dating during the week, as well as marriage were for-
bidden in many of these contracts while weekly church attendance and Sunday school teaching were often required.

That stereotype is gone today. It has been replaced

by an often militant unionist who is willing to take a place
on the picket line as well as to suffer the penalties inflicted
under laws forbidding strikes by public employees. As a
result the past ten years have seen an almost phenomenal
increase in the income of teachers, which, of course, has
rendered them ever more independent in regard to terms of
employment. This phenomenon has been met with great enthusiasm by members of organized labor who see it not only
as furnishing additional strength by virtue of the numbers of
teachers involved but also as a prestige situation because of
the status of the teacher especially in the eyes of that segment of society. On the other hand, the taxpayer who is
not a parent of school children is resentful of extra costs as
well as of reduction of teachers' obligations, for example,
through contractual limitations on class size. Parents are
often in a position of both supporting and opposing organizational efforts of teachers depending on whether they see the
particular activity at issue as beneficial or harmful to their
children.

A Single Major Cause for Dissatisfaction?

It would be naive to believe or to assert that there
is only one cause for all the militancy evidenced in the unrest surrounding American schools today. I take the position, however, that there is one overriding cause which
plays a like role among all other "causes." I refer to the
organizational structure of instructional processes in American schools. Instructional processes disregard to a substantial extent individual differences among learners, the
existence of which at all educational levels no one can or
does deny.

This is not an appropriate place to discuss the nature
of these individual differences but it safely can be said that
such differences are much more marked than is generally
assumed. Nor in this chapter will I dwell on the many efforts that have been made to adjust instruction to such differences. Suffice it to say here that unfortunately many of
these efforts have, in fact, served to accentuate rather than
to alleviate the failure to take individual differences into account in planning instructional procedures. Both of these
problems are adequately discussed elsewhere in this volume.

I visit many classrooms during the course of each
school year. During such visits I am impressed with how

little change has occurred since the days when I first attended
public school, before the beginning of World War I with regard to adjustment of class organization and to accomodation
of individual differences.

        With few exceptions, all schools today are based on
the same inflexible graded organization that existed during
my childhood.   With only slight variations, because of differences in the wording of state laws, all children above a certain designated age (usually about six) are required to attend
school.   When they first appear there are many obvious differences between them in addition to those connected with
their physical size and general appearance.   Some of them
are much more mature than others.   Some have had previous
"school" experience in kindergarten, in nursery schools, or
in Head Start.   The experiences that the several children
have had in this respect vary widely from child to child as
to their nature, quality, and intensity.   Other children come
to school without any organized pre-school experience but
obviously all children have had educational experiences of one
kind or another regardless of whether they have previously
attended classes of any kind.   Some of these experiences
have been broadening, confidence-building ones but others
have been very narrow and extremely stultifying.   Some
children have never in all their lives up to that moment of
entering school been more than a few blocks from there;
others, such as "Army brats" for example, will already have
travelled to and in foreign lands.   Some of these boys and
girls come from loving homes while others have experienced
indifference or even active rejection.   The homes they come
from also vary on the economic scale; some of them are rat-
infested slums while others are luxurious establishments.
Some of the children in one way or another have spent much
time with their peers while others have literally never played
with or even associated with other children.   Among the
school entrants are those with older siblings who have thoroughly briefed them about what to expect.   Along with these
youngsters come those who are only-children or who are the
senior siblings in their families.   Some are healthy, others
frail and highly susceptible to infection.   Some children put
up a bold front regardless of whether or not they are, in
truth, frightened by this experience while their colleagues do
not conceal their feelings and weep copious tears.   Some
mothers bring their boy or girl on the first day of school
while other children have to manage this first confrontation
alone.

There is also a significant difference in the degree of
readiness for learning among these youngsters.   Some will
have already learned quite a bit about reading while at the
other extreme there are those who literally have never handled a book and in many cases do not even know that there
is such an activity as "reading. "

To completely describe the range of differences existing among these entering first graders would take much time
and a great deal of space.   I will therefore merely state
that the foregoing is not by any means meant to constitute
the major part of a complete listing.   All these children,
however, are placed in the first grade and during the next
nine or ten months they will be expected to acquire certain
skills which have been pre-determined to be appropriate,
probably many years before any of the children were born.
They will be expected to have acquired certain reading skills,
some knowledge, most often of a rote nature, of "numbers"
and so on.

These foreordained tasks are completely independent
of the qualities and developmental readiness of any particular
child.   Every single child is expected to acquire these certain abilities by the end of the school year.   In the event that
he has mastered these skills and abilities before entering
school and does not augment them during the school year he
probably will not be penalized in any way.   The reader may
recall the experience of "Scout" in To Kill a Mockingbird.
Although she had already learned to read fluently before entering school, Scout had to partake of reading readiness
work.   Likewise, the effort expended by an individual during
the first school year is of little concern compared to whether
or not he has acquired mastery of subject matter to the degree expected of all first graders.

I am well aware that what I have written will be regarded by some readers as either unfair or exaggerated.
In truth it is neither.   The requirements at this stage in the
formal educational process, of course, are not unduly oner-
ous and most teachers tend to humanize the system as much
as they can.   As a result most children--in fact, nearly all
children--are barely aware of these conditions.   The fact,
nevertheless, is that they exist.

I do not in any sense join in the chorus of criticisms
of the modern school in which every negative situation is
generalized into a universal and then made the basis for

condemning everything about all schools and all individuals associated with them.   I am the first to want to pay tribute to the sincerity of the many members of school staffs, both teachers and administrators who have done so much to create more palatable conditions in their schools.   Nevertheless I sincerely believe that I have described a situation that is found in the majority of schools in the United States.

Being "Held Back"

The overwhelming majority of the entering first grade class is promoted to the second grade at the end of the first year and to the third at the end of the second year and so on.   However there are children who are not promoted.   If a child has failed to acquire or to come near to acquiring one of the skills that has been designated as a prerequisite for promotion to the second grade, he is often "held back." For example, even if he is well ahead of his group in understanding mathematical concepts, but had not developed any readiness for reading instruction, he may be required during the following year not only to repeat his previous experiences in reading readiness but also his first year's experiences in arithmetic.

During his second round in the first grade he must now spend his days with a new group of children learning the same songs, again being told about ways of remembering to bring his handkerchief to school daily, hearing for the second time the arrangements about feeding the classroom goldfish and, despite his full understanding of first grade "number work," once again using representative materials such as beads and chips to learn the numbers from one to five.

Although he is noticeably larger than his new classmates, our little friend has no conception how difficult his second year may be.   He will soon learn.   His playmate of last year has found new friends in the second grade and much prefers not to walk to school with someone who is a "first grade baby."   At morning recess he is pointed to by his former classmates but is rebuffed by them when he eagerly runs over to greet them.   When guests come to his home in one way or another his repetition of the first grade always works itself into the conversation.   Even his teacher, who seemed so patient last year, now is busy with the new first graders and when she calls on him is intolerant of his mistakes and in front of the entire class chides him, telling him

that he should certainly know that answer because he had all
that last year. In fact, soon each day becomes more difficult than the one before. For a while, pretending to be ill
early in the morning solves his problem but very soon it is
discovered that his peculiar illness occurs only on school
days so he has to go anyway.

    Before leaving this youngster, it is interesting to
note what happens to him at the end of the second year in the
event that, as often occurs for reasons that seem rather obvious, he still cannot meet the requirements in reading.
That this often happens should be no surprise if one considers
that even when he tried he couldn't seem to put it all together
during the first year, so that when he finally gave up trying
during the second year he still could not put it all together.
There are various ways in which school systems cope with
this situation. Some will hold him back for yet another year
with even more disastrous results. More commonly, there
is a rule about more than one retention at any one level, so
he now comes into the second grade, even more poorly
equipped to cope with the tasks of that grade than he would
have been a year ago. In some instances he may be required
to attend special summer session school classes before being
promoted. In such instances, however, the promotion is
most often associated with his attendance record rather than
with his scholastic improvement during the summer.

    In concluding this case, let me emphasize one point.
A search through educational research literature reveals not
one shred of evidence that this process of retention accom-
plishes anything positive whatsoever either for the child involved or for the creation of a greater homogeneity at any
subsequent grade level. The same, incidentally, holds true
for the process of acceleration. We cannot go into the matter of "skipping" a grade with adequate thoroughness here but
it does seem appropriate at least to call attention to the utter
inconsistency of the reasoning employed in setting up a graded
system and then allowing this process of acceleration to take
place.

    The fundamental basis of the graded system is the be-
lief that there is an almost fore-ordained special order or
sequence in which skills in any subject should be taught.
Accordingly, second grade skills, for example, cannot be
taught with any degree of efficiency to pupils who have not
first acquired mastery of the first grade skills we have been
discussing. Teaching the third grade skills successfully, in

turn, rests on the pupil's mastery of the second grade skills
and so on and on.

Given this justification of the graded school system
(and I can think of no other) how can the idea of acceleration
or the omission of the teaching of one whole block of prerequisite skills possibly be justified? The fact is that whatever
other difficulties skipping a grade may present, not having
had the skill instruction assigned to the previous grade does
not seem, in practice, to present even a minor obstacle to
the pupil's success in the next grade.

Intra-Class Grouping

Let us now turn to another point in the graded elementary school--for example, a fifth grade. Carefully performed studies as well as the observations of experienced
teachers tell us that by this time the range of individual differences in achievement in the various subject matter areas
has expanded greatly. Normally in a fifth grade, one might
expect to find a five-year range in the achievement of subjects like mathematics and reading. In other words, while
there will be some fifth graders who are able to cope with
the mathematics ordinarily assigned to the seventh grade,
there will be other children who can obtain only those scores
on standardized tests that would be considered norms for
second graders. Actually this range, which is usually "revealed" by the scores obtained on standardized tests, is
probably even greater because many of these tests base the
grade-level equivalents on the idea that the ceiling of a fifth-
grade test cannot be higher than the seventh grade. This
means that even a perfect score would be equivalent only to
a seventh-grade score. We are left to wonder how the student might have done on a test which included more difficult
questions. At the same time any score whatsoever (in the
case of one very widely used test, even writing one's name
on the test paper) will be regarded as being equivalent to the
second grade level.

We need to take a look at a typical fifth-grade class
to understand how individual differences among pupils are in
fact often completely ignored in instruction. We will not
dwell on the tens of thousands of fifth grades in American
schools where whole class instruction prevails, so that no
pretense is made that individual differences are taken into
account. Instead we will examine a class in which intra-class

grouping is employed so that individual differences may be accomodated.

In visiting fifth-grade classrooms we may be struck by two procedures which are often found. First, the number of groups formed is almost invariably three. Not two, not four, not any number other than three. Second, with rare exceptions each of these three groups is made up of one-third of the children in the class.

Now, if in fact, the intra-class grouping is organized to meet individual differences, it seems to me that the effort should be to establish groups that are as nearly homogeneous as possible. It would be almost incredible if in any class such a selection revealed three groups of identical size. In fact, most teachers admit that the establishment of three groups of equal size is a convenience in terms of classroom management rather than the result of a genuine effort to establish the most homogeneous groups possible. I will frankly concede that, as a proponent of individualized instruction, I find it impossible to believe that forming fewer than 30 groups in a class of 30 children will produce intra-group homogeneity in any case.

Even if one refuses to take what might be considered by some to be such an extreme view, it still would be unlikely that every class, year after year, would have a population that would inevitably fall into three groups. Even more doubtful is the idea that these groups would always be equal in size. It seems inevitable to me that some of the members of the highest achievement group would be much nearer to the second or middle group by any sort of measure or standard than they are to the very top students of the class. One does not have to be a statistician to have heard that a classroom group's scores or achievements tend to fall into the normal bell-shaped curve in which there is a piling up of scores in the middle with relatively few cases at the extremes.

It seems to me that the foregoing discussion raises substantial doubts about the efficacy of three equal-size groups in any classroom so far as the possibility of dividing the children into the most homogeneous groups is concerned. However, this scheme has an even more serious shortcoming as a means of accommodating individual differences between pupils.

According to my observations, when group procedures are used in a classroom for both reading and mathematics, often the same assignment to groups is used for both subjects. This implies, of course, that those children who are best in reading are also best in arithmetic and so on. Research studies have established beyond doubt that this simply is not so. Even more astounding is that the assignment of children to these groups is often not based, as one might expect, on either the mathematics or the reading achievement scores of the boys and girls but rather on intelligence quotients.

Three factors make such a criterion for partition of a class into groups more likely to result in heterogeneous than homogeneous groups: first, the margin of error on paper and pencil group tests of intelligence is so large as to render the results almost meaningless; second, at the very best an intelligence quotient is related to the capacity of an individual to learn rather than to the amount already learned; and third, the nature of the intelligence test, especially of a group test, is such that two persons may end up with precisely the same scores by correctly answering two different sets of items; that is to say, one person might answer correctly only those items dealing with verbal abilities, while another might answer an equal number of non-verbal items correctly.

Now if one grants that these groups into which boys and girls have been divided, with the very best of intentions, are in fact quite heterogeneous, it becomes interesting to look more closely at a typical group in a fifth-grade classroom of 27 pupils which has been divided into three groups.

Let us examine the "best" group: it may be called the "green group" by the teacher who naively believes the children will not catch on to the fact that she regards this group as superior to the others. As in each of the other two groups there are nine children in this best group. Of these nine children two individuals have IQs between 125 and 135. To the extent that the test scores on which the IQ index is based are at least partially valid and/or reliable, this means that these two children have highly superior capacities for learning. The other seven children in this "best" group have IQs between 108 and 102. (Incidentally the top child in the second or "orange" group also has an IQ of 102 but his name started with a letter further down in the alphabet than the child in the top group with an IQ of 102.) This, to the

extent that we can credit the test, means that these seven
children are quite average in their capacity for learning.

We observe that during the reading period all nine of
these children are given identical daily assignments in identical basal readers and in identical workbooks. They are all
expected to read the same number of pages (no more, no
less) in the reader and, after completing the assigned reading, to fill in the required answers in the blank spaces on
identical pages of the workbook. They will take from four
to six weeks to complete the basal reader even though the
top children in the group have already read the entire reader
on the first day that it was distributed, while the teacher was
not looking.

Are these children in the "green" group learning how
to read? Perhaps or perhaps not. In any event it seems
obvious that they are not learning even that with optimal
materials at an optimal rate. Are they learning to read?
I think the answer has to be, "Almost certainly not."

Now let us look at this same group during the arithmetic period. All three groups are given assignments: one
group will recite while the other two groups will be given
"seat-work" to do during the period. This assignment quite
often will consist of a large number of examples involving
the identical principle. Some children who have learned the
principle whether by rote or with understanding fly through
these similar examples. Other children who for one reason
or another are not sure of the principle involved have to
work out each example individually and fail to perceive the
common element in all parts of the assignment. Still other
children make the same type of error in each of the examples
because they perceive the similarity but not the nature of the
principle involved. The slowest or "beige" group is given
the same amount of work to do but the level is different for
each of the three groups. All pupils will, however, be required to finish their textbooks by the end of the year.

It is interesting to note that in many classes while the
in-class work may be differentiated, the homework is assigned to the class as a whole. Little thought is given to the
different amounts of time that will be involved in completing
these homework assignments or of the frustration experienced
by slow pupils who are never able to complete these assignments.

But let us return to our observation of the "green" group in the fifth-grade class we are visiting. The girl who has the IQ of 135 does not function at all well in arithmetic. From the day that she started going to school (practically every morning), she has had it dinned into her by her mother that arithmetic is very difficult and that she is bound to have trouble with it just as her mother--who now "can't add two and two"--had when she went to school. What one would expect to happen to the self-image of this 11-year old, who very well knows that she has a high IQ because the teacher told her mother who in turn told her, has in fact happened. She is always first in anything having to do with reading, social studies, and the like, but here she finds herself in a group where she is the only one who cannot do all her arithmetic work quickly or for that matter, at all. It is a dilemma at first but one that has been at least temporarily solved through an agreement with her "best friend" to copy the arithmetic answers from her paper.

Much more might be said about the relationship between the organization of present day classrooms and those of 60 years ago. However I shall merely add that Frederic Burk of the San Francisco State Normal School aptly described this approach at the turn of the century as a "lockstep," because this type of school organization is so rigid.

Learners at all levels have felt that they were not being treated as individuals but merely as part of an amorphous mass. It is my very firm belief that this feeling of not being treated as a single unique human being who has unduplicated needs and talents lies at the root of much of the dissatisfaction expressed and accounts for the remedies demanded by students and their parents all the way from the first grade to the university's graduate levels. There is a great push for relevancy--relevancy for the individual. There is strong protest against the enforcement of regulations directed at the entire learner population which are regarded as irrelevant to a particular individual. Indignation is expressed about teachers who do not know the learners in their charge as individuals. All of this supports the belief I have expressed just as it lends credence to my thesis that our present-day school organization is a major contributing cause of the present dissatisfaction with the educational establishment in the United States.

Resistance to Change

It should be emphasized, however, that there is a tremendous amount of opposition to the concept of substituting an individualized instructional approach for present procedures. Much of this opposition may be attributed to vested interests who have a great stake in the continuation of present procedures. There is also strong opposition from those who firmly believe that such an approach is not only impracticable and unworkable but also undesirable. These critics believe that individualized instruction would not produce satisfactory results. In short, they contend that children would not learn what they should learn at the time such learning should take place.

Because individualization of instruction is adequately described in the subsequent parts of this book, I shall not discuss its nature and techniques at this point. However, in order to understand the widespread skepticism and opposition to its adoption I suggest the five basic working assumptions on which this concept rests:

(1) Learning is promoted by giving the learner a part in determining what shall be learned, how it shall be learned, and when it shall be learned.
(2) The amount learned should be consonant with the ability, the motivation, and the interests of the learner.
(3) The pace of learning should also be suitable to the ability, motivation, pace, and interests of the learner.
(4) Evaluation should be based on the progress made by the individual learner rather than on a comparison with other learners or with arbitrary standards.
(5) Learning should never be an entirely solitary task so that there should be ample opportunity for sharing one's unique learning experiences with others.

It is obvious that these five basic assumptions contrast sharply with the implicit assumptions underlying the conventional approach. These may be stated as follows:

(1) There are certain fundamental skills and knowledges which must be mastered by everyone.
(2) There is one logical sequence into which these skills fall and the primary skills preceding any particular secondary skill must be mastered before that skill can be taught.
(3) Any kind of fundamental differentiation of the above

principles in the case of a particular individual is basically
undemocratic.

(4) Since the child when he matures will have to be
a part of many groups as he takes his place and role in society, group instruction is the most satisfactory way of
learning and constitutes the soundest preparation for life.

(5) Since we live in a competitive society, it is beneficial to evaluate a child's work in terms of the accomplishments and achievements of others.

It seems to me that these two sets of assumptions are
fundamentally irreconcilable.   One has to accept one or the
other and then adjust one's behavior accordingly.

It is true that the acceptance of individualization of
instruction would necessitate changes in the present structure
of education.   It is also true that there is always resistance
to changes in social structure.   On the other hand we live in
a changing world in which it seems to me changes are not
only desirable but absolutely necessary.   The computer today
does work in a few minutes that only a few years ago would
have taken one person a whole lifetime.   A mediocre come-
dian on a weekly television show is heard by more people
every single week than the total of all the people that heard
Abraham Lincoln in his lifetime.   Communication has become
so incredibly swift on a worldlike basis that it has outdated
traditional means of passing on information from one generation to another.

We must also be aware of the vast explosion in man's
knowledge.   It is no longer possible for any one human to
read "all about" a subject of any importance.   As we begin
to put books, reports, and research findings on microfilm
and on microfiche we soon realize that even this minimization
of space will not be sufficient for long.   In my lifetime the
recognition of the power of the atom and the exploration of
outer space have both come out of the age of Jules Verne
and into the reality of our lives.   How can we conceivably
reject teaching approaches merely because they represent
changes.   We can no longer afford the luxury of the easy
way out.

I am not a mathematician in any sense of the
but I address this book to the particular attention of
of mathematics especially at the elementary school
know from personal experience how very painful
ing group instruction in arithmetic can be becau

able to compete with the best in other subjects I was always behind in math. Just as I was almost at the point of understanding a particular concept, we left that topic and went on to the next. In the October 1970 issue of the Review of Educational Research there is a review article on attitudes toward mathematics by Professor Lewis R. Aiken of Guilford College. A careful reading of this article leads me to the conclusion that my frustrating experiences and my consequent dislike for and fear of anything mathematical, a feeling that lasted for many years, is shared by many others. I think that the most effective way of creating a change in attitude toward math is to be found in the adoption of an approach to teaching which is individualized. It is quite remarkable that in a number of studies that are reported in subsequent chapters of this volume a change of attitude is often associated with the adoption of the individualized approach even when for one reason or another there was no difference between the traditional and the individualized groups in test scores on the subject matter taught.

While I shall have the opportunity to mention it again in several places in this book, I wish to emphasize that individualization in and of itself is not a panacea which will solve every problem in the teaching of mathematics. No doubt there are some teachers and even perhaps some children who are psychologically oriented in such a way that individualization would be a very painful experience for them. There is no justification for requiring such people to participate in this approach.

We are at a point in time where American education is failing to accomplish its purposes for many young people whom it is charged to service. If we would but listen to these young people, if we would but realize that our society cannot continue to endure failure in the teaching of even the simplest fundamentals to many of our inner city children, if we as teachers search our hearts, then surely we must consider change. This necessary change may or may not include the adoption of the individualized approach to instruction but I do believe that this is a possible change worth looking into--hence this book.

II. INDIVIDUALIZATION OF INSTRUCTION

It is the purpose of this chapter, which contains four passages, to present some thought-provoking ideas about the rationale for individualized instruction and about ways by which this idea is actually implemented in practice. At this point our emphasis is on the aspect of "individualization" in general, rather than specifically on the teaching of mathematics. It would, I think, be a grave error to assume that everything about individualizing the teaching of a particular subject is unique. The fact is, there are certain common factors in the individualization of instruction that have universal application among all subjects, although obviously, there are also certain aspects that are unique to each special situation. It therefore behooves the student of individualized instruction first of all to direct his thinking to the common elements of all individualized instruction.

In planning the individualization of teaching and learning in any specific situation, one set of suggested criteria would include these postulates:

First, the primary emphasis must be on accommodating the multitude of individual differences that exist among the members of any group, each of whom is a unique individual with a set of characteristics not duplicated in the makeup of any other individual. It follows that to be successful, individualized teaching must be based on as thorough an understanding as possible of many aspects of each of those individuals who will play the role of learners. This is not to imply that the unique characteristics of the teacher should be overlooked in the planning.

Second, individualized learning can only be effec‍t   ‍e there is a degree of "self-selection" or choice on the the learners.

Third, there must exist a recognition of th‍

33

that not all children are going to learn the same quantities of material. To say along with the pioneers of the first quarter of the century, such as Preston Search [[1]end of article], Frederic Burk,[2] and Carleton Washburne,[3] that the essential thing is "pacing" is an unrealistic view in the light of what has been learned in the intervening years. What was implied by this early "individual" instruction, on which the individualized approach of today has been built, was that all children need to learn certain identical quantities of material but that there should be a difference in the time allowed each individual to master this assigned task. In practice today, this theory says that every fourth grader must master certain prescribed skills before he will be permitted to go on to the fifth grade (unless, of course, he has already "been held back" at this level a certain number of times). The fact that this idea has never worked out in real life apparently has not affected those who even now talk in terms that seem to attribute magic qualities to a list of skills that must be mastered (or else!) at certain stages of schooling.

And fourth, individualized instruction must never become isolated instruction. There must always be a provision for sharing with the group as a whole and for opportunities to work with others in temporary groups, small or large, made up of colleagues who have similar needs at the time.

There must be provision for evaluation that is carefully planned and soundly designed. The fact that proponents of individualization are inclined to see almost no value in total test scores, in the equation of like grade levels, or in intelligence quotients and the like does not mean that they are not concerned with ways of determining the degree to which the objectives of an individualized program have been realized. The diagnostic value of tests used to determine individual strengths and weaknesses rather than to arrive at a "magical number" has a major place in the thinking of those espousing individualization.

The four articles in this chapter all deal to some extent with theory, but the two by Spaulding and Grant are accounts of how individualization "works. " This aspect is one that is of primary concern to those who are first considering the idea of individualization. Therefore I begin this collection of readings with a selection which describes in considerable detail one plan used to individualize learning. I leave it to the reader to determine to what extent this particular plan and others to be examined later on meet the criteria set forth

above.   It is important to realize that there are many plans
for individualization just as there are many sets of criteria
that can be used in judging their merits.   No single plan or
program is a panacea; no single program would be desirable
in every situation.   The reader will, I hope, be concerned
more with the applicability of the plan described to the situation at hand than with its absolute merit.   Individualization
is emancipatory not only for children but also for teachers
and others who plan programs.   It is unlikely that any plan
described in this volume or elsewhere would be desirable or
even suitable without some adaptation in the classroom of the
particular school with which the reader may be associated.

Robert L. Spaulding

The author of the first article, Robert L. Spaulding,
earned his Ph. D. degree at Stanford University after a number of years of teaching and school administration in a variety of interesting elementary schools.   After a stint of college teaching and administering, the years between 1965 and
1970 were spent as a member of the staff of Duke University's Department of Education.   It was during this period
that plans were laid for the procedures described in the first
selection in this chapter.   In 1970 he returned to California
where he is now a professor of Education and director of the
Institute for Child Development and Family Studies at San
Jose State College where, if his past record is any kind of
indicator, he is engaged in novel, innovating, and exciting
programs.

Whatever the merits of the procedures described in
his article may be from the standpoint of the reader, it certainly cannot be said that this program was concocted from
inexperience or naïveté.

Archibald B. Shaw

The second selection, which presents concepts and
ideas rather than a program description, is also written by
one who has a rich and broad background in many aspects of
education.   Dr. Shaw is presently at Michigan State University.   I have no idea how he would react to my views on i
dividualization, but I am certain that when he wrote this
editorial he did not have in mind specifically this type ʳ
dividualization of instruction.   Nevertheless, it seems

that the principles and needs described in the passage lend strong support to the ideas presented in this book.

Jettye Fern Grant

        The third selection is by a practitioner of education, a teacher. It has been said that individualization is "a grassroots movement" because it is so often initiated by the classroom teacher rather than imposed from above by administrators or supervisors. It is for this reason also that individualized instruction is such an incredibly persistent phenomenon in education, because a teacher, in order to avoid utter frustration, inevitably seems to arrive at the realization of the need for individualizing, not only the rate of learning, but the quantity and nature of that learning as well.

        Dr. Grant's doctoral thesis, from which the third part of this chapter is extracted, is one of the most remarkable of the many hundreds of dissertations that I have read and one of a very few that makes this vast reading worthwhile. A common argument one hears against the employment of individualized instruction is that while it is "fine theory" it "won't work" in practice, in the school situation being discussed. So we can find in the literature on individualization of instruction the following kinds of statements:

        Individualization should not be used in the lower grades because children are not ready for it.
        Individualization should not be used in the upper grades because there is already ample provision for individual exploration there.
        Individualization should not be employed with highly intelligent children because they are already performing adequately and so why "quarrel with success?"
        Individualization should not be employed with dull children or with slow-learning children, because they could not cope with the independence involved.
        Individualization should not be used with small classes because there the teacher is in a position to give individual attention to children without such a program.
        Individualization should not be used in large classes because such an approach is not manageable with so many pupils.
        Individualization should not be employed with middle-class children because they already have the opportunity to do individual tasks in out-of-school life.

Individualization should not be used with disadvantaged children because these children need the psychological support given them by a more structured program.

Individualization should not be instituted with better than average school achievers because they are the successful products of the existing program so "why rock the boat?"

Individualization should not be employed with lower than average achievers because these children need the sequential structure of the traditional program.

Individualization should not be used by young, inexperienced teachers because they will have enough difficulties and problems without embarking on such a complex venture.

Individualization is not suitable in the classroom of the experienced teacher because such teachers have already learned how to make their presently employed approach an effective one or, even if perchance they have not, they will in all likelihood be unable to adapt to the radical change of philosophy required.

These propositions force one to think of the fable in which each mouse agreed that it was a splendid idea to bell the cat.

The fact is that individualization has succeeded at each of the levels, with each of the groups of pupils mentioned, and in every situation described in the list above. (This is not to say, however, that individualization has always succeeded whenever it has been tried.)

Most vehement are the objections that are voiced concerning the use of individualization in classes of disadvantaged children who have fallen behind in their schoolwork and who live in what is euphemistically called "a low socio-economic area"--slums, inner city, or ghetto--in substandard housing and under substandard conditions impinging on every single phase of their lives.

The second strongest objection is to the concept that individualization might be extended beyond a single subject.

Dr. Grant's thesis is a description of a program of individualized instruction in a class of 29 disadvantaged children who, by the time they were promoted to her fourth grade, were well behind the standards set for children completing the third grade.

Dr. Grant describes in her thesis the individualization

of not one subject area but of the entire curriculum, not for three or four months or even for one school year, but for a three-year period.

The excerpts taken from this thesis are largely concerned with arithmetic. It would have been equally justifiable to include a passage dealing with any other area of the curriculum or of school living because at this point we are more interested in the concept of individualization than in the teaching of a particular subject.

I cannot urge too strongly that those readers who are as fascinated by the description of the work of Dr. Grant as I am will find it exceedingly rewarding to read her entire dissertation. A microfilm copy can be purchased from University Microfilms of Ann Arbor, Michigan for a nominal sum. Photocopies can also be obtained from the same source but are considerably more expensive.

Another remarkable aspect of Dr. Grant's work is that after finishing her dissertation, she took two other groups through similar three-year cycles which were, of course, modified and improved in respect to both the procedures and the materials used, because of the knowledge acquired from prior experience.

I do not make any comment about the nature of the individualized program itself, as it is my purpose to leave such matters to you, the readers.

J. Murray Lee

The fourth and last item in this chapter is written by Professor J. Murray Lee of Southern Illinois University, where he was chairman of the Elementary Education Department from 1958 to 1968. Before that he was dean of education at Washington State University for 13 years. He is best known as co-author of Lee and Lee's The Child and His Curriculum, published in 1940 and revised in 1950 and 1960. This textbook has been used by uncounted thousands of future teachers throughout this country. He is also co-author of the Lee-Clark Reading Readiness Test.

Obviously Dr. Lee possesses the credentials to express his concepts about individualization of instruction. Here we find a skillful analysis of the need for individualized instruction as well as a discussion of the reasons for the failure of some teachers to carry on such individualization, even while recognizing the need for it.

While the reader may or may not agree with the ideas expressed in this article he certainly must grant the worth-whileness of carefully examining and considering them, because of the author's long experience in education.

In a recent letter, Professor Lee wrote as follows: "If you wanted a P. S. for the article here it is--'Since this article was written, there has been a marked increase in administrative arrangements and in materials of instruction which makes it much easier to individualize instruction'. "

Notes

1.   Preston Search was superintendent of schools in Pueblo, Colorado in the 1890's.   His work is described by him in "Individual Teaching" which appeared in the Educational Review 7:154-70, 1894.

2.   Carleton W. Washburne was superintendent of the Winnetka, Illinois schools from 1919 to 1925 and wrote extensively about the famous Winnetka Plan which he developed there.   He was one of Burk's students at San Francisco Normal School.   One of the best descriptions of his educational philosophy is found in his Adjusting the School to the Child: Practical First Steps published by the World Book Company in 1932.

3.   Frederic L. Burk was many years head of the San Francisco Normal School where he influenced many of the pioneers in the individualization of instruction.   He described his philosophy in forceful terms in Lock-Step Schooling and a Remedy which was published by California's State Printing Office in 1913.

PERSONALIZED EDUCATION IN SOUTHSIDE SCHOOL

Robert L. Spaulding

The concept of individualized instruction covers a wide range of efforts to tailor the educational program of the

schools to the individual learner--to his achievement and his
rate of development. The literature on child and adolescent
development shows that normal growth and development occur
in various rates from year to year and vary considerably
from child to child. Educational programs based on statistically derived patterns of average growth and development are
bound to be out of step with the developmental and growth
needs of most of the individuals in any given group of children. Longitudinal studies have shown that a great many
different patterns of development regularly occur among children. The intricacies of individualized patterns of growth
and response to the school environment make it impossible
for any teacher to plan and prescribe a program of instruction that would be completely appropriate to the individual
educational needs of each child in her class.

Even though it may never be possible for a teacher
to plan and prescribe a school program that would allow for
maximal learning by each child in her class, many teachers
have found it feasible to move away from a uniform instructional program, to which all children must adjust, to a curriculum that takes some account of the various rates of development and levels of achievement in their classes.

These attempts by teachers to meet the individual
needs of children have taken many forms, and new administrative instructional arrangements are continually being in-
vented. Among the most well known have been ability grouping, interest grouping, achievement grouping, individual contracts, tutoring, emergent curriculums, unit teaching, project or activity programs, and programmed instruction.

One basic problem underlies all the efforts to modify
the educational program of the public schools to meet the
needs of individual children. This is the problem of adequate
diagnosis of needs at each point in the educational development of an individual child. A closely associated problem is
the problem of treatment. Not only must diagnosis be accurate, but treatments appropriate to each of the learning
problems presented by children must be worked out. Those
who have watched the field testing of Individually Prescribed
Instruction (IPI) developed by Robert Glaser, Joseph Lipson,
and others, of the Pittsburgh Research and Development
Center, are aware of the tremendous complexities involved
in diagnosing individual cognitive performance in a structured
area such as mathematics or language and in prescribing appropriate educational encounters for individual learners.

Even if accurate diagnosis and appropriate assignment of learning experiences are possible in highly structured fields such as mathematics, language, or science, there still remain problems in working out school management systems that will use programmed materials keyed to diagnostic or evaluative measures. Without seeking to convey the idea that efforts for diagnosis and prescription of appropriate learning experiences should in any way be curtailed, I feel that alternative approaches should be pursued. They may prove to be more fruitful and less expensive.

Underlying the development of most continuous progress plans, as well as the IPI system, is the assumption that the teacher or a teacher's aide or some other representative of the school will remain in a dependent relationship to the authorities of the school who administer the diagnostic and prescriptive procedures.

The concept of teacher control over the selection and the sequencing of the curriculum presented in the schools is inherent in most programs of individualized instruction. The reluctance of school authorities to share control over the educational experiences that are provided students underlies much of the current unrest among students in our schools and colleges. What is relevant cognitively to one learner is irrelevant to another; and no teacher, however well trained and however experienced, will ever be able to anticipate how any particular program of experience provided by the school will transform the thinking of each individual learner. The demand by students to share in the decision-making process in the educational enterprise stems mainly from the frustration they experience as they are required to participate in meaningless, trivial, or redundant curriculum offerings.

Participation

Any innovation that purports to meet the needs of individual students in American education today must take into account the increasing demand that educational experiences be relevant. To assure relevance, all teachers must be past masters at accurate diagnosis of individual levels of achievement, rates of development, cultural backgrounds and personal patterns of cognitive development. Teachers must also be aware of educational alternatives so that the choices they make will be eminently appropriate for individual learners.

Students themselves have provided a clue. Their argument in a word is "participation." If one listens to students these days, their argument boils down to the assertion that only the learner himself can experience the relevancy of his own environmental encounters. If one grants that the students' assertion is valid, what is the role of the school? How much should students share in decision-making about curriculum experiences? Within what limits and with regard to what goals should decision-making by students be permitted? How might school organizations be restructured to permit greater freedom, responsibility, and decision-making by students?

The answers to these questions are not readily available, and only after experimentation with new systems of educational organization and instructional programming will tentative answers be found. The evidence is clear that answers are needed. Students demand more freedom, want more responsibility, and insist on their right to participate in decision-making that affects their educational opportunities, occupations, careers, and sense of well-being.

An Experiment

A program of individualized instruction that permits continuous progress with increasing degrees of freedom, responsibility, and decision-making is underway at the Southside School in Durham, North Carolina as part of a five-year experimental program for disadvantaged children. The program at the Southside School is part of the Durham Education Improvement Program, a Ford Foundation funded study of the developmental patterns of about 200 children growing up in low-income settings. The study is examining the effects of various types of educational intervention.

One intervention at the Southside School is called "Personalized Educational Programming." It is designed to personalize the educational program to the extent that each child participates in decision-making concerning his daily educational program with increasing degrees of freedom and responsibility.

The "Personalized Educational Programming" experiment at Southside School is based on educational concepts developed during the 1920's and the 1930's by Carleton Washburne in Winnetka, Illinois and Helen Parkhurst in Dalton,

Massachusetts. The Dalton School in New York City continues to operate under the Dalton Plan. In both the Winnetka
and the Dalton plans, students were expected to undertake assignments or contracts, and work at them at their own rates
for as long as necessary to complete them within an allotted
or contractual period. Individual learners were given increasing responsibility as they demonstrated the ability to
complete contracts on time or to meet deadlines for specific
assignments.

In the Dalton Plan, as operated in New York City during the 1950's, students were given monthly assignments that
were uniform for a specific grade level. Since the Dalton
School enrolled children from high-income families, uniform
assignments based on grade-level standards were given.
Where individual differences in pupil ability made it unlikely
that a particular child would be successful in completing an
assignment, teachers were given the authority to lower their
expectations and modify the assignments for that child so that
he could continue to participate with the class as a whole.
However, most parents found the slowly developing pupils
needed to be tutored (out of school) to keep them from falling
behind the school's achievement standards.

In a series of adaptations of the basic Dalton Plan
tried out by the author in a number of public schools at the
third-, fourth-, seventh-, and eighth-grade levels, it was
found that the uniform assignment system was inappropriate
for most public school situations. A modified Dalton Plan
was worked out at the Hopland Elementary School in Hopland,
California. Under the plan, students participated in choosing
activities and tasks appropriate to individual levels of achievement and areas of interest. In the Hopland program the
contractual idea, taken from Carleton Washburne's work,
was incorporated into the basic Dalton Plan. The Hopland
Plan proved feasible in an ungraded seventh- and eighth-
grade combination class of 30 children in which the level of
achievement varied from Grade 2 through Grade 13. The
pupils in the Hopland class were from unemployed families
on an Indian reservation, from farm families, and from
business and professional families in the community. The
Hopland program was designed to be conducted in a self-contained classroom by one teacher, but it provided the basic
rationale and the structure of "Personalized Education Programming" at Southside School, which enrolled about 60 children aged six, seven, and eight.

The Southside Plan

　　　The 60 boys and girls were grouped into four family
groups, or "prides," which met from 8:30 a.m. to 9:00 a.m.
for planning and met again from time to time during the day
as a group whenever a group activity, such as a field trip
or a physical education activity, warranted. A pride was
made up of children from each of the three age groups represented in the primary school and was carefully composed
to form a heterogeneous group. The children in each pride
were chosen on the basis of academic achievement, rate of
learning, sex, race, and degree of socialization. In the
first year of the pilot program two prides were composed of
non-readers, and two prides were composed of children who
had some skill in reading. This accommodation to ability
grouping was made as a transitional step to the current year,
when each pride has representatives from the full range of
abilities.

　　　From 8:30 until 9:00, each child plans his daily
schedule with the assistance of his pride teacher or a teacher's aide. To complete his plan for the day, the child examines several posted schedules and pays attention to specific
requirements or constraints. Each day the pride teacher
lists on a conference schedule the conferences that each child
is expected to attend during the day. Some of these she will
conduct; others will be the responsibility of other teachers in
the school, but each child is expected to examine the conference schedule and put down on his daily plan opposite the
appropriate time the location and the subject of the conference. Once he has examined the conference schedule and
placed the conferences on his daily plan, he is ready to begin to plan his open laboratory time. Constraints on his
freedom are listed on the wall, on a chart rack, or at the
bottom of his daily plan sheet. In general, these constraints
impose on him the requirement that he spend at least a half
hour each on mathematics, reading, writing, and spelling
practice and an additional half hour in a project activity in
either social studies, art, music or science. Some of the
scheduled conferences will satisfy requirements in one or
another of these academic areas. If so, the child may omit
that area from his planning. When he has completed his
daily plan, all the activities on the required list are to be
included somewhere in his plan.

　　　Lists of possible ways in which a child may satisfy
time requirements in mathematics, reading, writing, or

spelling practice are posted. Suggested projects in social
studies, art, music, and science are also listed, but a child
may derive his own projects from social studies or science
units that teachers introduced to him during group conferences in social studies or science. In fact, teachers encourage pupils to pick individual projects that will complement the unit of group study in social studies or science.

In completing a daily plan each child examines the
"work station schedule. " This schedule lists the times that
each classroom or laboratory is open and the number of
children permitted to sign up for each of the stations at any
given hour. In designing his daily schedule, the pupil picks
a work station for each period during the day and inserts
his name opposite the work station, provided that the maximum number allowed at the station has not been reached.
He continues to pick activities and work stations, placing
them on his daily plan sheet until he has filled in all the
open time slots. He will, of course, include time for going
to lunch, physical education, and other routine activities.

When a pupil completes his daily plan, he presents it
to his pride teacher or a teacher aide for approval. The
signature of either one indicates that the plan had been approved. After the plan is cleared, the pupil begins his
activities for the day. Throughout the day he is expected to
have his daily plan sheet with him. All the teachers in the
program are encouraged to examine the daily plans of any
child who comes to a work station and to place on the plans
a symbol, such as a star or a set of initials, to indicate to
the child and to the pride teacher that the pupil has arrived
at a particular station on schedule. Similar symbols are
placed on the daily plans to indicate display of appropriate
study habits, work underway at the proper time, or completed on schedule. Positive comments for quality work,
creative ideas, or products, or developing skills are also
entered on the plan sheets.

Allowances for Differences

Since decision-making and planning are complex processes that must be learned, children are given small degrees of freedom at the beginning of the program in the fall.
The first two or three weeks teachers or aides plan the children's programs, much as in a conventional, teacher-directed
program. After pupils become familiar with the work stations

and with the "daily plan," children who show the ability to
read time and to govern themselves in a responsible manner
are given a half-hour for an appropriate activity of their
choice and a place to complete the activity. After a child
works effectively for a week or so with one half-hour slot
available for choice-making, additional half-hour periods are
made available gradually--consistent with each child's ability
to govern himself and to operate with greater freedom. In
all cases, however, the child's choices are constrained by
the guidelines posted in his pride and listed on his plan
sheet. In addition, each pride teacher holds routine weekly
or biweekly conferences with individual pupils to review their
daily plans (for the past week or two) and to make recom-
mendations on how their plans and their performance might
be improved. In some cases some freedom (a few time
slots) may be taken away for a week or two until performance is improved. In other cases increased freedom may be
awarded. The pride teachers keep records of individual pupil conferences, and the records form the basis of parent
conferences as well as judgments regarding specific instructional programs needed for individuals and small groups of
children in the prides.

     Reporting to parents and to pupils is based on the
daily plan sheets and the progress records kept by the teachers. The progress records accumulated by the teacher
through weekly or biweekly conferences are supplemented by
standardized tests and other evaluative instruments associated
with programmed materials.

     When individual pupils have progressed to the point
where they can plan for more than one day at a time, a
weekly plan sheet is used to let pupils work out projects that
will take several days to complete. Most pupils in the middle grades will eventually use weekly plans or even monthly
plans. When pupils are using weekly or monthly plans, it
is possible to incorporate out-of-school activities into the
over-all instructional program as individual pupils see the
relevance of out-of-school activities to their in-school projects.

     "Personalized Educational Programming" at Southside
School is extremely flexible and permits the involvement of
supplementary equipment, materials, resource centers, and
specialized personnel. To enrich the curriculum offerings,
children may, during laboratory time, use tape recorders,
headsets, phonograph records, games, a variety of

projectors--filmstrip, cartridge, slide, 8 mm. , and over-
head--and other materials and automated equipment.   Special
conferences are scheduled to introduce the mechanics of
operating new equipment; then individual children are instructed and tested in the use of the equipment.   A list is
made of pupils who are free to choose to operate the equipment on their own during laboratory time.   Other, less
skilled children are assisted by an aide.

A variety of programmed materials such as reading
laboratories and Individually Prescribed Instruction mathematics are made available to the children in the resource
center.   Special teachers who have one or two hours a week
available are included in the program.   Their available time
slots are listed in the work station schedule.   Teachers assign some of the children to the special teachers, other children who are more reliable schedule themselves during lab
time.   When new teachers or new equipment are introduced
into the school, pride teachers set up special instructional
conferences to discuss the new laboratory opportunities and
to demonstrate the new equipment.

Balance

Since social skills, concepts, and interpersonal relations are best learned through group activity, it is important
that pride teachers plan the social studies program carefully
to involve all the children in the school.   Many individualized
programs tend to isolate children from one another and reduce the amount of peer interaction and group learning.   In
"Personalized Educational Programming, " pride teachers
plan units of study in the social studies, which become a
major source of group goals and activities.   Out of the units
of study come many small group and individual projects that
are carried on in laboratory time, but the over-all goals of
the units are set up by the teachers.   Many academic skills
are practiced in social studies projects, but academic skills
are also taught in special instructional conferences on each
of the skill areas.

Summary

After one year of experience in the pilot program
most children were able to handle a substantial part of their
own educational activities responsibility.   Many of the

children who are in our disadvantaged group were unreliable
at first and needed constant supervision and direct instruction, but all but one or two have made significant progress.
The greatest problems occurred where noise and movement
in one work area interfered with instruction in adjacent work
stations.   The physical layout of the school in 1969-1970
separates verbal instructional stations from construction, art,
and dramatic play areas.   Distracting movement and noise
were major deterrents to effective communication in the instructional centers.   With the current physical arrangement
of instructional areas and work stations, many of the problems of interference can be overcome.

        The future of individualized instruction in the public
schools depends upon the ability of school authorities to develop systems which will transfer greater and greater decision-making power to children within carefully structured
limits that guide pupils and help them learn ways to govern
themselves and achieve on their own.   The problem of socialization in a technological society has become increasingly
complex as mobility has increased and patterns of employment have been transformed.   New educational systems are
urgently needed to prepare all children to govern themselves
and assume greater responsibility in planning their own educational experiences.   Any system of education that is designed to be controlled completely by school authorities,
however individualized it may be, seems to this observer to
be incompatible with the economic and social imperatives of
modern, technological America.

        Students are demanding relevancy in all their educa-
tional experiences, and none of our educational institutions is
exempt from increasing pressure to abandon instructional and
administrative patterns that are no longer appropriate to the
educational needs of students.

        The innovative program at the Southside School in
Durham, North Carolina is one of the many possible arrangements that offer increased degrees of freedom and decision-
making within gradually broadened limits.   Learning to use
freedom responsibly is fundamental to the effective acquisition and use of knowledge in a democracy.

INDIVIDUALIZED INSTRUCTION

Archibald B. Shaw

Of all the tough never-ending problems that face
school and university people, the problem of making education fit the individual is one of the hardest.   Yet few issues
find educators and the American people so closely in agreement.   People are different.   We cherish their individuality.
Yet our institutions are basically designed to teach groups
and are largely oriented to common needs.

The problem is twofold.   The ways individuals learn
vary, and the kinds of curricula that challenge them and
serve them best are widely different.   Most teachers make
honest efforts to meet the first challenge, although some just
don't know how.   But as youngsters progress and curriculum
gets institutionalized, the provision for genuinely differing
needs dwindles down to a few tracks complete with social
and status distinctions.

Because reading, writing, listening, and speaking are
fundamental skills required of substantially everyone, the major adaptation to individuals must be made in the methods of
teaching and the kinds of learning experiences confronting
children.   The conscientious third-grade teacher can testify
how difficult and demanding a task that can be, even if she
is fortunate enough to confront only 25 eight-year-olds each
fall.   The job is further complicated by the social distinction
of high-achievers and the despairing self-image of the youngsters with slower or different development patterns.

As youngsters grow older, differences increase and
multiply, partly through the inexorable operation of growth
formulas (IQ 80 and IQ 120 are three years of "normal"
growth apart at age eight, but five years aparty by age 12),
but perhaps as significantly through the different effects of
regular success for one and persistent failure for another.

The kind of individualization that will maximally develop needed common skills for every child is largely the
challenge of the teacher.   The administrator who is sensitive
to the problem can only support and encourage the process.

To one view, the effort to obliterate differences by ability grouping is often a futile contribution.   Every child differs significantly in how he learns best, and any grouping which overlooks differences may aggravate the problem.

Perhaps the most subtle but pervasive pressure comes from the confusion between the ever-clearer universal need for general or liberal education and the definition of what constitutes liberal education.   The postponement of technical-vocational courses to post-high-school years and the increasing pressure for a solid four-year liberal arts education as a prerequisite for professional study are symptoms of this confusion.

Let us pass over the elite-building selection and reward process, and let us even accept the Platonic idea of a hierarchy of occupations.   But what, in the name of good sense and humanity, are we (the elite decision-makers) going to do with the millions of youngsters for whom cajoling and browbeating, and even individually paced instruction, just won't work?

Maybe we will ultimately find out how to enlist and retain in our programs these kids whom God must love, He made so many of them.   In the meantime, a million or more drop out of the procession each year, before they can develop employable skills and certainly before they have achieved the generative benefits of general education.   What can we do?

To one view, the problem calls for all the ingenuity, adaptability, and statesmanship we can muster to find and establish curricula which are truly geared to the individual.

A LONGITUDINAL VIEW OF INDIVIDUALIZED INSTRUCTION

Jettye Fern Grant

Educators have become increasingly concerned with the problem of dealing with the great range of abilities, interests, and needs among the pupils in any elementary classroom.   Much has been said about the need for a program of instruction which enables each child to progress at his maximum rate, and in his unique way, toward the full realization of his potential ability.   The vast accumulation of evidence

on individual differences indicates a need for such procedures
as the individualization of instruction to a degree not commonly practiced in classrooms in the United States.

The individualization of instruction is, in fact, a major area of study and experimentation in many schools throughout the country at the present time. The investigations reported deal mainly with the use of (1) grouping arrangements,
such as team teaching, ability grouping, and ungraded
classes; (2) specialized equipment, such as teaching machines
and language laboratories; (3) programmed learning sequences,
multiple textbooks, and other instructional materials on many
subjects and at many levels of difficulty; and (4) new curriculum content and new methods of instruction. In each instance the use of such materials and devices has been introduced primarily for the purpose of facilitating learning by the
individual.

Most of the attempts which have been made to work
out some effective method of individualizing instruction have
been focused on the teaching of reading. Very little research
has been done on the individualization of instruction in other
areas of the elementary curriculum. No study has been reported on the individualization of instruction in all subjects
within a single class and no reports are available of longitudinal experiments with individualization.

Recognizing the limitations which have characterized
experiments in individualized instruction, the investigator attempted to push beyond current boundaries and provide for
the pupils in one class a continuous three-year program of
individualized instruction through the intermediate grades
four, five, and six, with appropriate differentiation of subject-matter content, instructional materials, teaching procedures, and learning experiences in all areas of the curriculum.

The principal participants involved in this study were
the teacher and 43 children in an elementary school in Berkeley, California. Of these 43 pupils, 22 remained for the
three-year period from the beginning of the experiment in
grade four through its conclusion at the end of grade six,
and are designated in this report as Experimental Group A.

For comparative purposes, two other groups also were
included in this investigation. One of these, known as Experimental Group B, was closely associated with Experimental

Group A.   The other served as the control group for the
study and is designated as Control Group C.

The children in these groups lived in one of the lower
socio-economic areas of the district; they attended an elementary school with a total enrollment of more than 800 students in grades K-6, inclusive.   The pupil personnel of the
local school was predominantly non-Caucasian in contract to
a predominantly Caucasian school population in the district
as a whole.

Operational Policy

It had been anticipated that the development and implementation of an experimental program of instruction would
involve the process of decision-making at every step, and
that the final outcome would depend, in large measure, upon
the quality of the decisions which were made.   It was deemed
necessary, therefore, to establish some guide principles
which would be given prior consideration in every decision-
making situation.   These principles served as operational
guidelines and helped to keep procedures consistent with the
purposes of the experiment.

(1)   In order to ensure a close working relationship
with the families of all the pupils, and to take advantage of
the diversity of their cultural heritage, direct two-way communication with the parents and other immediate relatives
was established at the outset, and frequent opportunities for
their involvement in the planning and operation of the instructional program were provided.

(2)   All persons directly concerned in each decision-
making situation had an opportunity to participate actively in
making the decision.

(3)   Only those materials and methods which were
deemed educationally defensible were used in the classroom.

(4)   The procedure which held the greatest promise
of good, first to all children in the class, and then to the
individual child, received the highest priority.

(5)   The major objective of the instructional program
was to teach the children how to learn by themselves: how
to recognize their instructional needs and to utilize all

available resources in meeting these needs.   Concomitantly,
the chief measure of the success of the program was the degree to which the children were able to free themselves from
dependence upon the teacher and to assume responsibility for
self-direction in their learning activities, both in and out of
school.

(6)   A systematic effort was made to get the pupils in
the class to cooperate in the establishment of a learning environment wherein the individual pupil would be not only permitted, but also encouraged and helped in every way possible,
to work at his own pace toward his optimum level of attainment.   The advancement of no single child was dependent
upon the achievement of any other; instead, every individual
began study at the place he had reached in each subject-matter sequence and progressed as fast and as far as he was
able during the time that he was enrolled.

(7)   A corresponding effort was made to develop within
the classroom an emotional climate in which each child felt
a personal concern for the individual rights and feelings as
well as the educational progress of his classmates.

(8)   Children were helped to establish and to maintain
within the classroom a democratic form of student-govern-
ment, with authority for the management of their affairs
delegated to them in an ever-increasing measure commensurate with their growing sense of responsibility and skill in
self-control.

Individualizing Arithmetic

Two pupils had failed to score on the test on easy
addition facts and needed to be retested.   One child had
demonstrated mastery of the facts at adult level and was
ready to be tested on harder combinations.   The other pupils
needed to study the facts they had missed on the test.

As an initial step toward individualized learning in
arithmetic, the pupils needed to discover that various arithmetic activities could be carried on in the classroom simultaneously.   While the children studying specific combinations
worked with semi-concrete materials on individual flannel
boards, the child who was ready for a harder test learned
how to take it alone by listening to the test record through
earphones.   This left the teacher free to work with the two

children who had failed to score on the test on easy addition facts.

The cause of their failure was not clear. Some possible reasons were: (1) They did not know the combinations; (2) they were reluctant or unable to respond to a new type of test; (3) they could not write fast enough to record the sums in the time allowed. In order to test further their knowledge of these combinations, the teacher gave each child a set of ten cards, 2" x 3" in size, on which were written the numerals 0, 1, 2, 3, 4, 5, 6, 7, 8, and 9, respectively. With these cards displayed in sequential order on his desk, each child was asked to listen once more to the easy addition facts as they were dictated at slow speed. But this time, instead of writing the answer to each fact, the child had only to touch the card which had the proper numeral. Either the mechanics of the test or the combinations continued to baffle these two children, and therefore a different method of assessing their knowledge of the combinations was devised.

Each of the children was given a set of 12 one-inch wooden cubes, or blocks. Then the teacher had them demonstrate their understanding of number words by responding informally to such directions as, "Show me three blocks; show me six blocks; show me one block." This they could do. The teacher asked them next to use their blocks to show the answers to the easy addition combinations by placing, as quickly as possible, the proper number of blocks on a colored sheet of paper. They went through several practice exercises, showing, for example, that two blocks were the sum of one block and one block; four blocks were the sum of two blocks and two blocks; three blocks were the sum of two blocks and one block. Then they started the real test and worked through a series of 30 combinations.

The results of this test were determined not so much from the child's ability to display the correct number of blocks as in the way he picked up the blocks, and the speed and confidence with which he displayed them. If, in response to the combination, three and three, the child picked up three blocks and then another three blocks, there still was no assurance that he knew the combination. But if he hesitated, tapping a finger, nodding his head, or making some other movement which indicated that he was counting up the total before picking up the blocks, it was quite clear that he did not know the answer. Only in those instances when the child moved quickly and confidently to pick up and display the appropriate number of blocks did the teacher credit him with a knowledge of the combination.

The final scores made by the children on this special test were recorded on the progress chart, and special notations regarding their particular difficulties were added to their individual records.

Arithmetic

Before entering grade four, the pupils in the class had been grouped for arithmetic instruction as follows:

No. of Pupils	Arithmetic Level	Page
11	Book 4	1
9	Book 3	195
5	Book 3	161
4	Book 3	130

On a standardized arithmetic test administered early in grade four, the class achieved a mean grade equivalent of 3.4; this was seven months below the grade standard of 4.1 for that date. The achievement of individual pupils ranged from a grade equivalent of 2.0 to a grade equivalent of 4.1.

The state-adopted textbooks which the pupils had used in grade three could not be adapted easily to individualized instruction; nor did they include certain material needed for a "newer" approach to the teaching of arithmetic. The teacher decided, therefore, to use the State textbooks for basic instruction, as required by law, and to use supplementary arithmetic textbooks to meet those needs for which the State-adopted series was clearly inadequate.

The Seeing Through Arithmetic Program, published by Scott, Foreman, was selected for use as the major supplement to the adopted arithmetic program. The sequence of instruction provided by this program was used in Experimental Group A, and is an example of a fully-detailed sequence which served as a guide (a) while instructional materials were being selected, (b) while long-range plans were being developed, (c) while the strengths and weaknesses of individual pupils were being assessed, and (d) while academic achievement was being evaluated.

The "four-step teaching method" used for introducing
new processes made the textbooks in the <u>Seeing Through
Arithmetic</u> series well suited to individualized instruction.
In Step 1, called "See," the pupil was shown in detail what
to do. A series of pictures showed the situation, the objects involved, and the action taking place. Then the process
with the symbols was shown and related to the visual repre-
sentation. Brief, easy-to-read verbal explanations were included. In Step 2, called "Think," another example was
worked out. Questions which helped the pupil to think about
the example and to notice important details were provided.
In Step 3, called "Try," other completely worked-out examples were given, but the pupil was expected to work these
out for himself and then compare his work with the completed
examples in his book. In Step 4, called "Do," examples for
practice were given. Also included in Step 4 were additional
examples for pupils who required more practice in order to
learn the process.

All material in the <u>Seeing Through Arithmetic Program</u>
was organized under nine headings:

<u>Learning how</u>	lessons using the four-step teaching method to develop or review computational skills
<u>Exploring problems</u>	lessons developing specific problem-solving techniques
<u>Moving forward</u>	lessons developing other new material, including use of ratios to express rates and comparisons, fractions, and measurement
<u>Using arithmetic</u>	maintenance and practice in solving verbal problems
<u>Keeping skillful</u>	maintenance and practice in computation with whole numbers and fractions
<u>Checking up</u>	tests of various types: inventory, achievement, diagnostic, and problem-solving
<u>Thinking straight</u>	lessons involving special aspects of quantitative thinking
<u>Looking back</u>	lessons involving reteaching or review of previously taught skills

Side trip   lessons on various topics re-
                                     lated to the arithmetic pro-
                                     gram; such as Roman numer-
                                     als and time zones.

The nature of this material enabled individual pupils
to progress, on their own, through a comprehensive, balanced program of learning experiences in arithmetic.

Arithmetic instruction during the early part of grade
three was pointed toward two objectives: (1) to help the children learn the basic arithmetic facts; and (2) to help them
develop a background of essential concepts and skills which
would enable them to use successfully the materials in the
Seeing Through Arithmetic Program.

Arithmetic phonograph records were an invaluable aid
in teaching the 390 basic arithmetic facts.   The arithmetic
facts were separated into groups corresponding to the drills
on the records, and duplicated on "study sheets. " Wooden
cubes, counting boards, and other manipulative devices were
used to teach the meaning of the combinations.   For the children who experienced difficulty in remembering certain combinations, matching games were designed to promote recall.
When a child believed that he had learned a group of facts,
he was permitted to check his mastery of the combinations
by playing the corresponding test record.   Very soon it became apparent that some children were taking the tests without having studied the combinations enough to learn them.
This practice created a problem because it tied up the phonograph and the test records needlessly.   Each child was requested, therefore, to pass a preliminary check and file a
verification before using the equipment for a test.   The
preliminary checking procedure was very simple.   The child
who wished to be checked filled out the first three lines of
a check form, then invited a parent, a classmate or a friend
to take his study sheet and test him orally on the combinations.   If he passed this screening test, he obtained the signature of the checker.   Then he asked two other individuals
to verify his readiness for an "official" test.   When he had
secured the signatures of three persons on the verification
form, he filed it with the teacher and took the test on the
record.   This is the form used to verify readiness for a
test:

Name_____ Date_____
I am ready for this test:
_____
                            Name of test
Record No._____ Side_____ Speed_____
Checkers: 1._____
          2._____
          3._____

A Class Progress Chart listing the names of the children in a vertical column, and the names of the individual
tests on a horizontal line provided spaces for recording each
child's performance on the respective tests.

The job of developing mastery of the basic arithmetic
facts, although time-consuming, was less difficult than the
task of providing an experiential background which would enable the pupils to proceed with the study of equations and
other unfamiliar aspects of arithmetic included in the Seeing
Through Arithmetic Program. Firsthand experiences and
concrete materials were used to illustrate the situations and
operations involved in different types of problems. First, the
children learned to dramatize a story situation. Then they
learned to verbalize the situation, the action, and the results.
Finally, they learned to show what was happening with pictures,
semiconcrete representations, and arithmetic symbols.

During this early stage of the experiment, the original
groups were maintained, but the divisions between groups
were flexible. Learning experiences provided for one group
often appealed to pupils in another group and could be understood by them, because they dealt with life situations common to all children in the class. Any child was free to join
another group for any lesson if he could show that it was a
profitable use of his time. Pupils in the more advanced
groups often joined slower groups to clarify concepts and/or
to review material previously covered. On the other hand,
children in the slower groups often joined pupils in the advanced groups because they could understand the concepts as
well as anyone; their difficulties usually related to the use of
symbols, not ideas. For instance, one child knew few of the
"arithmetic combinations," and was unable to add or sub-
tract; yet he sold newspapers in the business district each
evening, and was very skillful in making change. He could
figure costs, estimate probable income, and calculate his
profits to the penny. Several children, by sitting in with
various groups, were able to progress rapidly from one group
to another, so that they were able, by the time they were

ready to use the new textbooks, to work at a higher level than would have been possible otherwise.   In most cases the children knew whether the work was too hard or too easy for them.   They knew whether they understood the material presented in a lesson, or whether they needed further instruction.   They knew when they were able to move ahead of the group and work independently.

It took much time for some pupils to develop the feeling of security which was necessary before they could discuss their learning problems frankly with the teacher.   But when this had been accomplished, the prospect of personal achievement provided a powerful incentive for study and practice in arithmetic.

The arithmetic skills and abilities of the pupils were regularly checked through the use of manipulative devices, pictorial situations, personal demonstrations, teacher-made tests, and standardized tests.   As soon as a child qualified for work in the supplementary arithmetic program, he was given a textbook and permitted to proceed independently.   By the first of December, in grade four, each child had begun to work in a supplementary arithmetic book.

The use of the textbooks necessitated the development of new instructional procedures.   As each pupil moved ahead at his own pace, it was needful to keep track of every individual: to know where, in the sequence, he was working; to know how well he was learning the material that he was covering.   Individual progress was recorded daily in an arithmetic notebook.   One page in the notebook was assigned to each child; his name was written in the upper right-hand corner.   At the end of the arithmetic period, the teacher called the "roll" as she turned the pages of the notebook. Each individual responded by telling the teacher (1) the level of his textbook, (2) the page on which he had been working, (3) the number or letter of the last problem which he had worked successfully, and (4) whether he was having any difficulty.   For example, a child might report, "Book 4, page 15, problem 2-H; no difficulty. "   Another child might report, "Book 4, page 43, section on Exploring Problems; I do not understand it. "   Whenever a child reported any type of difficulty, the teacher immediately set a time when she could work with him on his problem.   A large portion of the teacher's time during the arithmetic period was used for this purpose.

It was not necessary, however, for a child to wait until the end of the arithmetic period to ask for help. It was his responsibility to obtain help whenever he needed it. If the teacher was busy, and another pupil could help him, he was expected to ask the other child for help. If no other child could provide the assistance needed, he was expected to write his name on a special section of the chalkboard which was reserved for emergency requests. As soon as the teacher saw a child's name written there, she checked to see what was needed, and arranged to provide the help desired at the earliest opportunity.

Teaching guides for the supplementary textbooks, with answers to the problems, were kept in a special bookcase near the teacher's desk. When a child completed a unit in his textbook, he was free to get a copy of the teaching guide and check his work. If all problems and examples were correct, the child returned the guide to the bookcase and proceeded to the next unit. If no more than two errors were found, and the child was able to correct them, he did so. Then he proceeded to the next unit. If he made more than two errors, he was expected to report to the teacher for help. Pupils were not permitted to refer to the answers in the teaching guide during the course of their computation. However, as their skill in reading increased, many children were able to read and understand the instructional procedures suggested for the teacher, and thus were able to work out most of the units for themselves.

The teacher scheduled a large part of the arithmetic period for individual help; however, these individual conferences often turned into group conferences. As soon as she determined the nature of a child's difficulty and prepared to help him, she announced to the class, or to other individuals who might be concerned, what she was going to do. Anyone in the class who wanted or needed that kind of instruction was invited to come for the lesson.

Some arithmetic periods were reserved for reviews, for testing, and for the presentation of new material. New problem types always were presented in simplified form so that they could be understood by every child. Then each child was encouraged to design similar problems which he could solve at his own level. When the more advanced pupils comprehended a new problem type, they were able to create other problem situations of the same type which involved more difficult computation. Pupils working at a common

level often competed to see who could design and solve the
most difficult problem.

As enthusiasm for arithmetic instruction increased,
the children spent more time on voluntary arithmetic homework. Their parents were familiar with the superficial aspects of the program, but they did not feel qualified to help
their children with problems involving equations, or sets, or
the newer methods of long division. A special course in
arithmetic instruction was provided, and most of the parents
and guardians attended the full series of meetings. There-
after, the parents and children often worked arithmetic together.

As the children advanced from one level to another,
additional supplementary materials were used. A great variety of manipulative devices, such as scales, speedometers,
abacuses, fraction kits, and protractors were used by all
children. A miscellaneous collection of arithmetic games
and puzzles was available to all the pupils, and used by
some. The Ginn Arithmetic Enrichment Program was provided for those who could use it successfully. Eight pupils
used Enlarging Mathematical Ideas (Book 5) and six also used
Extending Mathematical Ideas (Book 6). These materials
were satisfactory for pupils who could do abstract thinking;
but they were frustrating to those who had difficulty in using
number symbols and in seeing relationships between numbers.

A special file of arithmetic practice materials was
prepared for the pupils in Experimental Group A. Any child
who needed problems for practice or review on any level,
and on any topic, could obtain them from this file. The
work sheets were coded, and the answers were provided on
separate cards.

In grade six the use of study contracts and a flexible
schedule made it possible for each child to schedule his
class work to suit his interests and needs. In arithmetic,
more than in any other subject, the children tended to arrange their schedules so that they could work together.
Children who were competing with one another checked frequently to see who was ahead. If one had been absent, he
often worked long hours at night to catch up with his friends.
It was a common practice for children working close together in the sequence to compare their answers before checking
the answers in the teacher's guide. If they found a difference
they tried to discover the reason for the difference

"on their own. "

The ease with which the program progressed may
have caused undue complacency on the part of the teacher.
There were no serious problems. The materials were quite
satisfactory, and required no modification for individualized
instruction. The children liked arithmetic. Four pupils completed not only the entire program in the state-adopted textbooks and in the Seeing Through Arithmetic Program textbooks for grade six, but also the seventh-grade textbook,
Seeing Through Mathematics 1. In 2. 8 years the pupils in
Experimental Group A achieved a mean gain of 3. 5 in arithmetic. The grade equivalents of individual scores on the
final standardized test ranged from 4. 4 to 10. 1. The mean
grade equivalent was 6. 9.

It is important to note that the pupils in Experimental
Group B and Control Group C achieved nearly as well. These
groups did not have the same program of individualized instruction, but they did have some of the same instructional
materials.

Outcomes

The problem of the study was twofold:

To develop a three-year program of individualized instruction for the pupils of one class; and
To evaluate the effectiveness of this program (a) by
comparing the achievement of the pupils in the experimental
group with the achievement of the pupils in one other experimental group and in a control group, as measured by standardized tests in the language arts, arithmetic, science, and
social studies, and (b) by comparing the change in achievement of pupils in the experimental group with the established
norms for the tests. This also served to test a null hypothesis: that the achievement of pupils in the individualized
program would not differ significantly from the achievement
of other children of comparable age and ability in the same
school, or from the national norm.

Experimental Group A was made up of 13 girls and 9
boys, with a mean IQ of 103, and a mean chronological age
of 9 years, 2 months at the beginning of grade four.   Experimental Group B was made up of 7 girls and 12 boys, with
a mean IQ of 104, and a mean chronological age of 9 years,

FIGURE ONE

RESULTS OF ACHIEVEMENT TESTS IN ARITHMETIC
AND READING ADMINISTERED TO PUPILS IN
EXPERIMENTAL GROUP A

From October 1960 through June 1963

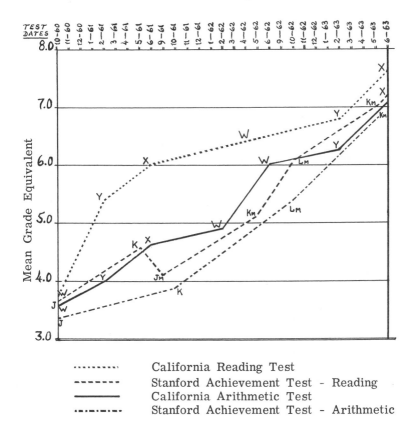

.... California Reading Test
‑ ‑ ‑ ‑ ‑ ‑ Stanford Achievement Test - Reading
————— California Arithmetic Test
.‑..‑.‑..‑. Stanford Achievement Test - Arithmetic

3 months at the beginning of grade four.   Control Group C
included 19 girls and 27 boys, with a mean IQ of 102, and
a mean chronological age of 9 years, 3 months at the beginning of grade four.

Experimental Groups A and B were located in adjoin-
ing bungalows situated in a remote area of the elementary
school playground.   Experimental Group A remained with the
same teacher through grades four, five, and six.   Experimental Group B had two teachers during the intermediate
grades; one teacher taught the class in grade four, and
another teacher taught the class through grades five and six.
Some of the materials and methods designed or adapted for
use by Group A were used by Group B.

The pupils in Control Group C attended classes in the
main building of the elementary school.   These pupils had
different teachers each year in grades four, five, and six.
They followed the regular school program and were grouped
within their classes for instructional purposes.   None of the
materials and methods designed especially for the individualized program were used by Group C.

The instructional program provided for pupils in Experimental Group A had the following important characteristics:

(1)   The pupils managed themselves through a system
of self-government.
(2)   The pupils learned how to learn by themselves;
that is, they learned how to identify instructional needs, to
work out a study plan, to follow through on the study plan,
and to check results.
(3)   The pupils helped to set their own goals through
the development of individual study contracts.
(4)   Each child progressed individually; that is, he
advanced as fast as he was able, without being pushed or
held back to keep pace with other children.
(5)   Each child operated on a flexible daily schedule
which enabled him to take up the day's learning tasks in the
order of their importance to him, and to complete each one
before moving on to another task.
(6)   The involvement of the pupils' parents in the development of the instructional program insured their cooperation and assistance in all class projects.

(7)   The use of a School Resource Volunteer to read

and discuss with the pupils their original stories resulted in increased interest and skill in creative writing.

(8) The use of reading job cards helped the pupils to establish purposes for independent reading, and helped to insure the development of basic reading skills and abilities.

(9) The provision of a daily newspaper for each child and a liberal supply of popular magazines, catalogs, and other current reading materials for the class helped to extend the children's reading interests and increase their knowledge of world affairs.

(10) The use of the tape recorder and the record player enabled the teacher to provide instruction for pupils working on different levels at the same time.

(11) The use of other equipment, such as the typewriters and the sewing machine, stimulated the children's interest in school activities and provided opportunities for new learning experiences.

(12) A program of High Expectations enabled each child to take "school" home with him.

Conclusion

(1) The effectiveness of the three-year program of individualized instruction is verified by the results of standardized achievement tests in paragraph meaning, word meaning, spelling, language, arithmetic reasoning, arithmetic computation, science, and social studies.

(a) The gains made by Experimental Group A were greater than the gains made by Experimental Group B in language. The difference was statistically significant at the .01 level. The difference in the gains made by Experimental Group A and Experimental Group B in other subjects were not statistically significant.

(b) The gains made by Experimental Group A were greater than the gains made by Control Group C in paragraph meaning, in word meaning, and in language. The differences in the gains in paragraph meaning and word meaning were significant at the .05 level. The difference in the gains in language was significant at the .01 level. The difference in the gains made by Experimental Group A and Control Group C in other subjects tested were not statistically significant.

(c) The gains made by Experimental Group A exceeded the standard gains indicated by the norms in all subjects tested.

(2) There is insufficient evidence to reject completely the null hypothesis: that the achievement of pupils in the individualized program would not differ significantly from the achievement of other children of comparable age and ability in the same school, or from the norm.

(3) Because there were many innovations in the individualized instructional program provided for Experimental Group A, it is impossible to identify the specific factors which caused Experimental Group A to make greater gains than Experimental Group B and Control Group C in language; and it is impossible to identify the factors which caused Experimental Group A to make greater gains than Control Group C in the reading tests of paragraph meaning and word meaning. The investigator believes that these results may have been due to her personal interest in these subjects, and her efforts in searching for materials and procedures which would insure the optimum progress of each pupil in the language arts. Perhaps it is reasonable to infer that equal diligence in the development of individualized techniques in all areas of the curriculum might have made a difference in the gains in other subjects.

(4) It is the belief of the investigator that the intangible benefits of the individualized program were more important to the pupils than the measurable gains in academic achievement.
(a) The children learned to value human rights and human resources.
(b) They learned to live democratically, to accept responsibility, to protect their freedoms.
(c) They learned how to make decisions and how to follow through with effective action.
(d) They developed self-confidence, and a sense of pride in their own accomplishments.
(e) They learned how to learn by themselves.
(f) They raised their expectations.

(5) The benefits which accrued to the investigator were important, also.
(a) She learned new ways of working with parents and pupils.
(b) She gained new insights into the educative

process.

(c) She developed some new and promising instructional materials and procedures which were intended to help each child achieve his full potential.

The writer offers to other individuals interested in
initiating some form of individualized instruction the following
suggestions:

Teacher and Pupils

(1) The teacher should begin to individualize instruction in the area of the curriculum with which he is most
familiar.

(2) The teacher may utilize one or more techniques
for individualizing instruction without making other changes
in the educational program or in the classroom organization.
For example, the use of job cards in reading or the adoption
of contract spelling would not necessarily affect other aspects
of the instructional program.

(3) Elementary school pupils should remain with the
same teacher at least two years. This helps to insure continuity of learning and provides more time for the teacher to
become acquainted with the needs and the potential of each
child. It also gives the teacher more leeway in planning a
program geared to the particular requirements of individual
pupils, for he does not have to be concerned about "getting
them ready for another teacher. "

(4) Pupils should be grouped into classes by chronological age, but not by ability. An ideal class for individualized instruction includes children with the normal range of
interests, abilities, and needs. These differences give pupils an opportunity to develop a tolerance and respect for
personal differences, and provide appropriate classroom situations in which to practice democratic living.

(5) Pupils should be encouraged to help one another
in school; the child who receives the help benefits from it,
and the child who gives the help learns as he teaches.

(6) Pupils should be given experiences in self-directed
learning. Children who are taught only to "do what they are
told" lose the ability to think what to do for themselves.

(7) Self-government should be instituted in the primary grades so that the pupils may learn as soon as possible
how to accept responsibility for their conduct and for their
independent learning activities. A functional system of self-
government also relieves the teacher of many duties related
to classroom management, and gives him more time to work

with individual pupils.

The Program

(1)   The elimination of competitive grading will help
the child to direct his attention toward learning what he needs
to know rather than to working for a grade.

(2)   The preparation of study contracts should be a
part of every child's instructional program.   Pupils work
harder to achieve educational goals which they have helped
to set for themselves.

(3)   A flexible daily schedule which is adaptable to a
variety of needs is more conducive to learning than a seg-
mented schedule which provides a specific amount of time
for each subject or learning activity during the day.   When
children become deeply interested in learning something, they
are reluctant to drop it and turn to another subject just because the schedule shows it is time for a change; and when
such a shift occurs, the children find it hard to give full attention to the new subject.   Large blocks of time are essential for many learning experiences.

(4)   A highly diversified program offers a greater opportunity for the development of each child's interests and
abilities than a program in which all pupils study the same
subject at the same time.

Instructional Materials, Sequences and Records

(1)   A great variety of equipment and materials on
many levels of difficulty should be available in the classroom.
A good classroom library is especially important.

(2)   Modern equipment, in good condition, enables the
teacher to utilize her full potential as an instructor.

(3)   A simplified system of record-keeping enables the
children to assume responsibility for maintaining many of
their progress records.   Fully delineated instructional sequences and abridged (or condensed) sequences and check
lists are essential for record-keeping.

(4)   Teacher's manuals and answer books which children may use to check their own responses enable the children to proceed independently, and save the teacher much
time.

Parents,   Community,   and School Resource Volunteers

(1)   The involvement of parents in a close working relationship with the school results in benefits for the child,
for his parents and the school as well.

(2)   Curriculum enrichment for disadvantaged pupils
is most likely to be a reality when it is a joint enterprise
supported by the home, the school, and the community.

(3)   The increased use of competent volunteer assistance will relieve teachers of many time-consuming tasks and
thus enable them to give more time to important instructional
matters.

INDIVIDUALIZED INSTRUCTION

J.   Murray Lee

To what extent does the school program need to be
individualized?   Should all children have the same school experiences?   What does a child need to know to be promoted?
These are questions discussed pro and con by many teachers.

There is a wide range of ability in each classroom.
Even in the first grade some children have begun to read,
some are ready to read, for some it will take nearly all
year and perhaps longer.   In a high school class you may
find some students who read on the level of college sopho-
mores or juniors.   Other students read on a fourth-grade
level, giving a ten- or eleven-year span of reading ability.
This range of reading ability is also approximated in practically all other abilities.

Each child has a wide range of individual abilities.
Studies have indicated that the difference in abilities within a
child may approximate about 80 per cent of the range of abilities within a class.   This means that each child has some
areas in which he does much better work than others.   He
has some areas in which he meets with considerable success,
other areas in which he has great difficulty.   Many of the
readers of this article would find it possible to read an average size novel in an evening, but would find the prospects of
a 20-mile hike uphill a devastating experience.

<u>Children come from different social and economic</u>
<u>backgrounds.</u> The extent that this statement is true varies
widely from community to community. However, even in a
community as homogeneous as Pullman, some children come
from homes in which the parents are just barely making a
living. Some come from homes of very rich wheat farmers.
Yet, it is felt that the town is fairly homogeneous. Contrast
this with another town in the state of Washington in which
children of white parents comprise approximately 73 per cent
of the school population. These children are divided between
children of transient workers, laborers, and a few professional people. There 16 per cent are children of Mexican
origin, 9 per cent of Indians and the remaining 2 per cent
consist of Filipino and Japanese children. The Mexican and
migrant white groups tend to live in low rent housing. This
housing consists of small one-room cabins which·are inadequate to properly house the customary large size family of
the migrant. The Indians live in unpainted, poorly-kept
houses.

Imagine the difficulty of the teacher with youngsters
coming from these many and varied home backgrounds with
different patterns of experience and cultural beliefs. The
difference in attitudes toward schools of the transient workers and the permanent citizens of the community is marked.
In contrast also is the attitude of the Mexican and Indian
parent with that of the Filipino and Japanese parent. We
know only too little about the effect of these differences upon
the needs of these children or upon the instructional program
which will be most effective with them.

<u>Children have various degrees of readiness for learn-</u>
<u>ing any one thing.</u> All teachers have heard about the prob-
lem of reading readiness. It is well known that some children are ready to read when they come to school, others
are not. The same problem exists in any phase of learning.
In high school chemistry some are ready to do the work in
chemistry for they have had well developed interests in science of long duration, while others have not the slightest
concept of what it is all about. Are these individuals equally
able to profit by the same program? What kind of individual
or group work needs to be done in the chemistry class to
meet the learning needs of each pupil? In too many classrooms this difference is disregarded entirely. The same
subject matter and experiences are given to all. There is
little or no attempt to individualize.

Children learn and grow at different rates.   The facts
of physical development are best known.   We know the rate
of growth.   Height varies from individual to individual.   We
know that some children talk before others.   The same
phenomena hold in all kinds of learning.   Children do not
learn at the same rate, nor can the same thing be expected
of each child.   However, when you look at classroom practices, this factor is too often disregarded entirely.   What
are the implications for individualizing instruction if this
phenomena is really recognized?

Children learn best when the learning task is within
their ability.   Faced with an impossible task children only
meet with failure.   Where children experience continuous
failure in the learning task, we realize there is relatively
little learning.   We know in the field of spelling, for instance,
where a poor speller is faced with 20 words a week in which
he is able to get only a few, that it is a poor spelling program.   He will actually learn a great deal more if he has
12 or 15 easier words in which he is successful in getting
them all.   He not only learns more spelling words, but his
attitude toward spelling is completely changed.

Children learn best when they feel secure.   Recent
research has indicated that this is a most important condition
of learning.   Insecurity creates blocks, slows up or prevents
thinking.   As the research findings accumulate, the emotional
climate of the classroom becomes more and more important.
Where there is an attitude that permits children to do their
best, where they know they are accepted and where they know
the teacher feels that they are important as an individual,
they learn better.

Children are motivated differently.   Motivation is the
result of both their background and their previous experience.
If effective learning is to take place, the child must be motivated to learn.   The pattern of his present motivation provides a base for beginning.   Unless this pattern is known
and understood by the teacher, she cannot capitalize upon it.
Here is an area in which each teacher will have to do more
to understand each child and to try to reach him.   The usual
recommendation, is of course, that the teacher begin with
his interests.   It needs to go further than this, the teacher
must also understand some of the motivations of the family
and the encouragements which the student receives from
home.

What do these eight principles add up to? They tell
us that all children cannot learn the same things in the same
way at the same rate.

Most teachers would subscribe to these eight principles
dealing with what we know about children.   The actual difficulty, however, comes in trying to interpret these into classroom procedures.   For instance, in a recent survey of the
schools of Utah, it was found that teachers believed in the
need for individualized instruction.   Yet, one of the weakest
areas was in their actual techniques of putting this belief into
practice.   Partly, of course, in large classes it becomes
very difficult to individualize or even work with groups.   However, much more can be done than has been done.

Let's take another instance.   In the field of spelling,
in too many classrooms throughout the nation, all the children in the room are taught the same words.   It has long
been a practice in reading in the elementary schools, to have
three reading groups, but seldom do we find three spelling
groups.   In a fourth-grade class, some youngsters might be
studying second-grade words, some third-, and some fourth-
grade.   There is every reason to believe that this would result in much more efficient learning and develop better
spellers.

In high school, there is a beginning of group and individualized work.   However, too often all children are given
the same reading assignment in history, or the same assignment in chemistry.   There is little attempt to vary the materials according to either their needs or their abilities.

Another indication of the fact that we believe but do
not practice, is in the worry which many teachers have of
whether youngsters should be promoted from the first grade
to the second grade.   In talking with many teachers this past
year this question constantly arose.   It seemed that the major worry was concerning the opinion of the second grade
teacher.   They know that for the children in question, it
would take them some time to develop into more effective
readers.   They know that the work should be adapted to their
level of presentability.   However, they still felt that while
they could individualize the work for the child, the second-
grade teacher would expect all the children to come up to
"second-grade standards. "

This situation is an indication that faculties need to

consider this problem of the continuous development of youngsters. They need to develop techniques for passing on information so that each teacher will know where to begin with
the child.

What Blocks Our Practice?

We do not adequately know our children. Much more
can be done in all schools in getting better acquainted with
the children, with their home conditions, with their abilities,
with their motivations.

We are not too conscious of how differences in background affect learning. Does a child who has travelled widely
have a better motivation for studying geography than one who
hasn't? Does a child who has interest in science have a
better motivation for studying science? What differences
exist in children from homes of lower social economic backgrounds in comparison with those from higher social economic
backgrounds? What differences exist in their ability to use
language, in their ability to understand various social concepts? There is some research in this area but the most
important thing is that each faculty study their own children
to see what differences exist in specific youngsters.

We do not have, nor do we utilize, the variety of
learning materials available. There is an increasing wealth
of materials to meet the needs of children of different abilities and different interests. Only a relatively few schools
are adequately equipped with these materials.

We do not utilize what we know about teaching small
groups or individuals. Probably more work has been done
in working with small groups in the field of reading than in
any other area. Some experimentation is going on at the
present time in developing individualized reading programs.
This has been very successful in some first grades. Groups
have also been used very effectively in social studies in the
elementary schools. There are other opportunities in arithmetic and in spelling. Perhaps here is an opportunity for
elementary and secondary teachers to work together to see
how group work can be efficiently developed at every level.

We still feel that all children should acquire a certain
minimum of subject matter. This concern is very clear when
you listen to teachers talk. They are not necessarily

concerned about stretching the abilities of the best pupils to their limit.   They do feel however, that all youngsters should have mastered a certain minimum amount of work supposed to be taught on that grade level.   If they fully accept the concept that children grow and learn at different rates, they would be willing to take the child where he is and provide for as many opportunities for his growth as possible.   This feeling of concern is also tied very closely to the following point.

We are afraid of what the next teacher will say. Teacher after teacher to whom I have talked has expressed a concern about what the teacher of the next grade will say when he or she receives the pupils.   They of course know that they are giving the best possible kinds of experiences to the youngster.

We can do many things as a staff to correct this situation.   We can improve and share our knowledge of children. We can pool our experiences.   Where a technique or procedure works for us it may help the next person.   Finally, we can pool beliefs and develop mutual trust.   The results of such a program will pay huge dividends for improved learning for boys and girls.   No longer will it be necessary to handle all boys and girls in exactly the same fashion or the same way.   Instruction can be carried on in small groups or individualized to meet the special needs.

III. A CLOSER LOOK AT INDIVIDUALIZED INSTRUCTION

We follow the rather general introduction to the concept of individualized instruction that was found in the last
chapter with more specific details in the five selections
placed in this chapter.

Purposely, we are still staying away from an emphasis on mathematics. The examples of individualization given
in this chapter run the gamut from individualized instruction
in the entire elementary school curriculum to individualized
teacher training at the college level.

I hope that a careful reading of these passages will
help to crystallize the feelings, attitudes, and interests of
the reader more efficiently than would be the case if we took
a direct plunge into descriptions and discussions of individualized mathematics instruction. I think that there are at least
three aspects of individualization that should be food for
thought after the careful reading of this chapter and the preceding one.

First, individualized instruction encompasses both
teaching and learning. It is of course impossible to ever
discuss teaching without considering learning but in the individualized process the union of these two activities is
strongly emphasized because it is so much more apparent
than in the traditional approach.

Second, individualized instruction is not an "alone
thing. " There are many more relationships with others than
there are in traditional whole class procedures. In the first
place, a completely new relationship springs up between
teacher and pupils. There are many times when a single
pupil has the full and undivided attention of the teacher--
something that seldom occurs in the traditional classroom.
The relationship between the learners is also altered in radical ways. The fact that everyone is engaged in more or less

different tasks and assignments leads to a new feeling of
partnership in a joint learning effort.   In a whole class operation some are often considered to be "dumb" by the others.
There is a patronizing air toward those in the lowest ability
group on the part of those in the highest group.   Under individualized instruction there is an atmosphere of mutual assistance and cooperation.   Quite possibly the poorest pupil
may be helping the brightest because the particular task requires a special skill (for example a mechanical task) that
the bright learner may not possess to the same degree as
the youngster who, in a grouping situation, would always be
assigned to the lowest group.   The tasks each child is working on are recognized as being different from, rather than
easier or harder than, the tasks of others.   The small ad
hoc groups that are formed and dissolved because of common
needs result in association of children who otherwise might
never learn to know each other well.

Third and finally, evaluation is on the basis of individual accomplishments in the light of individual capacity, individual experience, and individual effort.   Useless rote work
is not expected from anyone.   If one particular boy or girl
has mastery of one type of arithmetic task, he is not assigned such examples for homework.   By the same token, a
child who finds an arithmetic process exceedingly difficult is
not assigned more examples for homework than he can handle
during a reasonable period of time.   In this way evaluation
does not become either frustrating or anxiety producing.
Surely in the 1970's we should have advanced beyond the
stage where children at any level of school should ever become physically ill because of an impending test or examination.

Fischer and Fischer

The first passage in this chapter was written by Barbara Bree Fischer, then of the University Elementary School
at the University of California at Los Angeles and now at the
University of Massachusetts, and by Louis Fischer of San
Fernando Valley State College of Northridge, California.
The process described in this article, which originally appeared in the Elementary School Journal, is a non-graded
plan in which team teaching plays a role, but with primary
emphasis on individualization.   The reader will have to do a
certain amount of "filling in" as the article is quite short,
but I think that the general picture that emerges is one that

will interest all those wishing to learn more about the process of individualization.

<u>William R. Kramer</u>

The second article is made up of selected portions of a mimeographed curriculum document prepared by William R. Kramer, principal of the Borel Middle School, a school composed of students in grades six, seven, and eight located in San Mateo on the Peninsula a few miles south of San Francisco.

The succinct explanation of his interest in individualization sent me by Mr. Kramer seems interesting to me:

"My interests in individualized instruction began with hearing the ideas of Dwight Allen on behavioral objectives. I visited Duluth, Minnesota, spoke personally with Dr. Esbenson, and visited individualized instruction classrooms. Upon my return we began to implement a similar system at Borel. I have also visited project PLAN and have had some contact with the IPI system. "

I selected this passage for inclusion in this volume because I feel that the plan is typical of a great many approaches that are being made in the direction of individualization. I also feel that it presents the procedures to be employed in a particularly lucid manner, so that the reader can determine exactly what it is that was being planned at the Borel school.

There are two shortcomings that appear to me to be an integral part of the description of this plan. I have not abandoned my idea of letting the reader be his own judge as to the merits or demerits of each plan presented in this book. I only point out these matters for the reader's consideration, without in the slightest degree implying that these ideas should necessarily be accepted by the reader, unless he himself sees them as valid in the light of his own perceptions of the situation.

1. As I see it, there is a complete lack of evidence that this plan for individualization was considered by the teachers before its adoption. The chance for the success of such a plan is greatly diminished if teachers do not have a part in planning and a genuine voice in the decision about adopting a

departure from customary procedures.   There is an old saying about leading a horse to water that has application here.
Also, in the present-day climate it would be wise and would
increase the likelihood of success to have children take some
part in planning.   This applies also to parents and to members of the community.

       If circumstances were such that the involvement of
pupils, parents, and members of the community was impossible (a circumstance which would be very unlikely to occur
in the case of an administrator who had a sincerely enthusiastic approach), then there should still be a program for a
thorough orientation for members of these three groups.   It
has been said that one of the major reasons behind the difficulties schools have with budgets, building plans, etc. is the
lack of communication between the school officials and the
community.   If this is true, and I certainly believe it is,
every effort should be made to carry out activities and to
communicate in ways that are in accord with the principles
of sound public relations.   This is especially important when
there are to be departures from former practices with which
the community has more familiarity than it has with the newly
proposed procedures.

       Of course, it is quite possible that this criticism is
completely unjustified and that Mr. Kramer did in fact pursue the necessary procedures outlined but felt that information about this did not belong in the curriculum document
from which the passage in this chapter was taken.

       2.   The second factor I wish to draw attention to is
the lack of self-selection or choice by pupils in the plan, as
outlined here.   As I read this plan there are a set series of
contracts in each area of study.   Let us say, for example,
that there are a hundred contracts that could be completed in
Subject X.   This would mean that there would be between 30
and 35 contracts available per year on an average.

       Mr. Kramer makes it very clear in his document that
it is part of the plan that pupils will finish a varying number
of contracts each year, for he says that each pupil will start
the new school year at the point at which he left off at the
end of the preceding year.

       Let us now assume that we have a pupil whose aver-
age number of contracts completed per year is 20.   At the
end of the third year we may expect that he will have

completed 60 contracts.   Pupils who complete the entire sequence will have attained the same stage at the end of their
second year.

The question I pose for the reader's consideration
during and after his reading of the Kramer article is this:

Would it perhaps be more desirable if behavioral objectives were varied according to ability and other individually
differing factors, so that the slower pupil would get a view
of the entire sequence by the time he finished the eighth
grade?   Would this not be more desirable from the standpoint of likely success in the ninth grade?

J. Clair Morris

The next excerpt is taken from a doctoral thesis written at Brigham Young University in 1968.   This is the second
excerpt taken from a doctoral dissertation.   It is my feeling
that one of the most valuable untapped resources for shedding
light on present day problems in education is the unpublished
doctoral thesis.

Many of them, it is unfortunately true, have very little
to offer (this by the way seems to be true at most institutions granting doctorates), are unimaginative, and often report on carelessly performed, poorly organized work and
base their findings on ridiculously small samples, which are
subjected to an experimental factor for only very brief periods of time.   In many of these theses the writers show only
a minimal knowledge of the literature, present inadequate and
often also inaccurate bibliographies, and use statistical analyses that are, to say the least, of doubtful merit.

In contrast there are dissertations that show the result of earnest application and which make real contributions.
In my opinion the dissertation selected for inclusion in this
chapter falls into the latter class.

Certainly Dr. Morris' dissertation shows the results
of years of work and makes a very substantial contribution
to knowledge about individualized instruction.   Dr. Morris is
presently superintendent of the Iron County School District in
Cedar City, Utah.   He has worked with a variety of ways to
individualize instruction as a teacher, principal,

superintendent, and consultant since 1954.

The variety of ways in which individualized instruction is perceived and therefore the variety of ways in which the concept is implemented is well illustrated in this passage from Dr. Morris' writing. It is an account of a five-year period during which there was a genuine attempt to individualize instruction in a high school. Results from a standpoint of statistical significance were apparently not all that had been hoped for, but it is certain that the findings of this research and the descriptions of procedures employed give many clues to ways in which the procedures used could be made more effective by modification in certain respects.

Charles J. Gorman

The last passage in this chapter is a slightly abbreviated version of Professor Gorman's short article in the Journal of Research and Development in Education.

Professor Gorman is in the Department of Education at the University of Pittsburgh. His article is a capsuled description of the plan developed at Pittsburgh for the complete individualization of teacher training. Obviously it is somewhat ridiculous to lecture prospective teachers about the desirability of individualization while training them in a restricted and traditional manner which proceeds as if the assumption were that all the students preparing to become teachers are possessed of the same abilities and experience and thus require the same amount and kind of training.

The reader will find this article as well as the complete plan from which it was extracted both useful and illuminating in thinking about individualization.

TOWARD INDIVIDUALIZED INSTRUCTION

Barbara Bree Fischer and Louis Fischer

A variety of new ideas for individualizing instruction can be found on the American educational scene. Perhaps the best known are non-grading and team teaching. Non-grading is a plan for grouping pupils to promote continuous progress in learning. In the non-graded school, teachers group children for instruction according to specified criteria other than chronological age. Team teaching is a pattern of staff utilization designed to increase flexibility.

Efforts to individualize instruction are, in part, the result of the realization that equal educational opportunities do not imply identical treatment of children. Certainly, identical treatment is the simplest approach to a group of children. Obvious exceptions to this notion of equality have existed in the schools for many years in the special educational programs for the blind, the deaf, and other "handicapped" children.

Two developments--one scientific, the other philosophic--have helped shatter confidence in this simplistic notion of equality. Scientific findings have established the fact that individuals differ in ways that are educationally significant and that they learn in a wide variety of styles. Philosophic analysis has also established the principle that in education equality means not equal education, but equitable or fitting education. The expressions "appropriate education, " "equitable education, " or "fitting education" are less ambiguous than "equal education. "

The general rationale of individualized instruction and its grounding in science and philosophy have often been stated in the professional literature, but not enough attention has been given to descriptions of how these ideas are used in the classroom. A detailed description of all that goes into the formation of class groups and instructional subgroups in a class is beyond the scope of any paper. A partial description is offered here in the hope that others will find it useful in their deliberations. The descriptions offers some idea of the careful consideration given to a child's intellectual, physical, social, and emotional growth in deciding the most

promising educational placement for him.   The presentation
is based on the assumption that individualization of instruction, or the provision of appropriate or fitting education, requires careful placement of pupils into instruction groups and
careful consideration of what goes on in such groups.   Intelligent placement can facilitate the efficiency and the effectiveness of individualization of instruction.

        In this case study, a team of three teachers worked
with 75 pupils from nine through 12 years of age.   Initial
placement in the total team group for the school year was
determined after careful study of each child.   The crucial
considerations were the intellectual, social, and emotional
development of the child as he related to the teacher's teaching style and the learning opportunities offered by the com-
position of the peer group.   In letters sent out in September
each parent was told how the school had arrived at decisions
on his child's placement and program: "We have considered
the kind of teacher who best stimulates and propels his
learning, the group of students with whom he functions most
productively and the educational program which should be
prescribed. "

        Individual records and notes on observations made
during the first few weeks of school showed a great range in
the children's past experience, learning style, interests,
autonomy in learning, achievement, attitudes, and skills.   To
place pupils into instructional groups, the team decided to
consider each child's ability to "learn to learn, " his autonomy
as a learner as indicated by his use of the teacher as a
learning resource, and his involvement in the learning task.

        A number of behaviors were considered in placement.
These behaviors are clustered at two points on a continuum.
As Figure 1 shows, one point is the more dependent state
of operation in a group, and the other is the more independent
state.   Most of the children were between these two positions, and all were uneven in their own pattern of learner
autonomy.

        These dimensions seemed to have a fairly high rela-
tion to other grouping criteria such as achievement in social
studies and skills in research, discussion, and committee
work, but seemed to have little relation to intelligence quotient, age, or over-all reading ability.

        After each pupil's behavior had been studied and

MOST DEPENDENT BEHAVIOR IN GROUP

In Using Teacher as Learning Resource, the Child

is very dependent on teacher for direction and support in order to achieve.
Constantly asks "What do I do next?" "Is this all right?" Cannot set own realistic goals.

avoids contact with teacher.
Turns away, avoids eye contact, does not ask for help when needed.

Involvement: The Child

is not in control of his ability to focus on classroom activities.
Is easily distracted by extraneous stimuli, does not distinguish relevant from irrelevant activity or ideas.

never or seldom shows enthusiasm about, or interest in, task.
Rarely smiles, or volunteers ideas, or uses past experience, or brings in helpful materials.

demonstrates few expressions of thoughtful work.
Satisfied with few disjointed sentences in written work, does not ask relevant questions, contributes unrelated or inappropriate ideas in discussions and conferences.

MOST INDEPENDENT BEHAVIOR IN GROUP

In Using Teacher as Learning Resource, the Child

is independent.
Can set own goals, choose appropriate alternatives to accomplish goals, proceed systematically toward goal, evaluate honestly. Makes a judgment as to when the teacher is the appropriate resource in this process. When he needs help, often turns to several reasonable resources to solve problem without teacher's guidance.

Involvement: The Child

has a long attention span for classroom activity. Resists inappropriate distraction from learning.

shows enthusiasm and interest in learning task.
Often contributes, searches out relevant materials at home and library, relates past experience with new ideas.

evidences thoughtful work in most learning tasks.
Willing to edit and polish written work, takes time to prepare and practice oral presentations, asks relevant questions, contributes appropriate ideas and information. Can express orally, or in written form, the criteria for a good job.

Figure 1.   Children's Behavior at Two Points of a Continuum

described on the basis of his autonomy in learning, children with similar behavior were assigned to a cluster. There were ten clusters in all. The ten clusters, which had from two to 13 children each, were arranged to form three instructional groups. There were 16 children in Group A, 25 children in Group B, and 34 children in Group C.

Children in Group A were the most dependent in learning; however, they showed their dependence in many different and highly visible ways.

Children in Group B tended to be concentrated along the center of the continuum of dependency. In many respects these children, too, were different from one another and uneven within themselves in relation to various behaviors that demonstrated independence.

Children in Group C tended to be the most independent in their learning; the pupils in this group were ready to practice the behaviors at the independent end of the continuum.

Figure 2 shows the range in the three groups with respect to age, intelligence quotient, and reading score. If the original large group had been formed on the basis of age or reading or achievement or ability, placement of the children on the basis of other criteria would have been more difficult. A program designed to carry out specified instructional goals must accommodate a wider range of factors related to learning if we recognize the variability of these factors within each learner, as well as among groups of learners.

Figure 2

Range in Ages, Intelligence Quotients, and General Reading
Scores of 75 Children in Group A, Group B, and Group C

Range in:	Group A (16 pupils)	Group B (25 pupils)	Group C (34 pupils)
Age	9-12	9-12	9-12
Intelligence quotients	90-160	100-145	105-165
General reading score	8.0-14.0	9.0-14.0	10.5-16.0

After the three instructional groups had been formed, the members of the teaching team worked together to plan or prescribe a learning environment that would encourage autonomy in learning.   For Group A the team proposed to encourage each child to grow in the desired behaviors identified in Figure 1.   For this purpose, it seemed important that the learning environment include:

A small group.   In a small group each child would have increased opportunities to interact with the teacher and would be subjected to decreased pressures from peers.

A teacher who had a supportive style of teaching and would insure emotional and intellectual support, as needed by individuals.   The teacher's support was to decrease as the children grew in independence.

The tapping of the child's immediate interest whenever possible.

Short-term goals that offered highly predictable success.   These goals were determined by the group at first and later by individuals.

Immediate feedback of results where possible, frequent self-evaluation, conferences with the teacher on specific goals set by the pupil.

Concrete, vivid experiences whenever possible, moving from concrete to abstract.

A variety of activities within a period, the length of time for each activity increasing as needed.

Each child was to work in the learning environment prescribed for a specified time.   Throughout the interval the pupils were evaluated constantly to insure appropriate placement.

Some changes in grouping were based on criteria other than those used in the initial groupings.   Several examples illustrate how decisions for these placements were made. The examples also illustrate the flexibility of the team organization in the effort to place children into learning situations most appropriate for them.

Sharon in Group A had a history of school phobia, but she had advanced so rapidly toward independence that she was soon ready for the learning environment of Group B, and plans were made to change her grouping.   During the planning, a question was raised about the effect the move would have on her friend, Debbie, who had been an isolate before her friendship with Sharon.   On the basis of autonomy

in learning, Debbie definitely belonged in Group A, but her
social needs at this time seemed so important to her total
learning that she, too, was transferred to Group B with her
friend.

Follow-up evaluation showed that Sharon was doing
well in all areas except in giving her ideas in group situations; she still tended to withdraw and needed much support
from the teacher.   Debbie showed growth in all areas, con-
firming the hypothesis that social learnings were important
for her and greatly influenced her behavior in learning to
learn.

Stuart, nine, and Rod, 11, two active boys in Group
C, were having difficulties that had not been foreseen.   The
boys seemed overstimulated by this group.   They anxiously
tried to contribute all the time and were unable to do their
usual high quality thinking.   They appeared to be "spinning
in their tracks. "   After careful consideration, they were
moved to Group B.   At a later evaluation it was obvious that
both boys were more relaxed and were performing at higher
levels in groupwork (they assumed constructive leadership),
in thinking (they worked on higher levels of Bloom's Cognitive
Taxonomy), and in autonomy (they worked alone on projects
or together in productive rivalry).

Some children shared responsibility for their placement.   Martin asked to work with Mr. Lasker.   "I think
I'd do better in his group, " Martin told us.   He was placed
in Group B with Mr. Lasker with the understanding that he
and his teachers would evaluate the appropriateness of this
placement.   At a later evaluation Martin expressed satisfaction with his placement but showed no specific gains.   The
teaching team agreed that he did seem happier, but he
showed no growth in autonomy, though the change did lead
to important positive learnings for him.   He learned that
adults did pay attention to his feelings and his ideas about
his needs.   He learned that his goals were not changed by
a change in teachers, but remained primarily the same because they were tailored by and for him with the help of the
teacher.

A few changes in grouping were based on the criteria
established for grouping; other changes were based on prior-
ities for individual learning such as teacher-child relationships, peer relationships, learning style, out-of-school or
in-school pressure.   The majority of the children in the team

did not change groups. They were able to practice the desired behaviors and grow toward independence in learning in the group where they were first placed. The learning environments met the children's changing requirements for the 12-week period.

An observer, watching teachers instruct groups of children, is not likely to understand how much teacher effort and competence in diagnosis go into the formation of instructional groups. Even the most sophisticated observer must ask questions about criteria for grouping before he can offer intelligent comments about the individualization of instruction. What occurs in instruction is vital; but serious professional attention must also be focused on the placement of children into instructional groups.

THE BOREL INDIVIDUALIZED SYSTEM OF INSTRUCTION

William R. Kramer

A Capsule Description of ISI: Individualized instruction consists of planning and conducting with each student a program of studies that is tailored to his learning needs and his characteristics as a learner.

The course of study is defined and organized through a series of sequential learning contracts. Each contract contains a specific learning objective stated in terms of what the student must do to demonstrate accomplishment of the objective (performance objective). The contract indicates the learning materials and procedures required. The teacher serves more often as an educational consultant to each child rather than as an imparter of knowledge. The student initiates and completes each contract consulting with the teacher as needed. Each student must successfully complete one learning contract before he may progress to the next.

Entry into this type of instructional program by Borel Middle School represents a commitment to the need for finding even better ways to successfully educate all of our children.

Historically, attempts at achieving individualization

have resulted in development of techniques of classroom
groupings.   Homogeneous grouping, tracking, groups within
the classroom and lately team teaching and flexible scheduling are examples of such increasingly sophisticated techniques.

In spite of such dedicated efforts, in most cases we
have succeeded in simply reducing the span of differences
that children in each group represent.   Any group-directed,
teacher-led program seeks an effective compromise and is
ultimately faced with a challenge it can never fully meet--
that of tailoring an educational program to the unique learning needs of each and every child.

Individualization at last looms as a truly realistic
goal.   The discarding of false assumptions on the teacher-
learning process has given us the inspiration.   The new
technology has given us the tools to accomplish this demanding task.   The gap between the desire and the doing can now
begin to be bridged.

The ISI program represents a shift from a teacher-
centered to a student-centered concept.   The transition to a
program which may be geared to each child's unique instructional requirements with differing interests, rates and styles
of learning, and intellectual levels is a massive undertaking.

Instructional objectives must be re-examined and re-
stated so that student accomplishment of these objectives is
observable.   A wide variety of learning resources, far beyond the restrictions of the traditional textbook must be
found, selected, and created.   Effective instruments for pre-
and post-evaluation of instructional objectives must be developed.   Techniques designed to free teachers from non-
essential functions and thus provide time for the teacher
must be developed and refined.   A climate in which the student assumes a substantial responsibility for his learning
progress must be achieved.

What Will Result From an ISI Program?

. . . . . a continuous progress (non-graded) curriculum
          which will encourage learning success
. . . . . growth in student acceptance of responsibility
. . . . . development of attitudes and study skills that will
          prepare a youngster for independent learning after
          formal schooling ends

A Group-Directed Classroom

. . . . . The bell rings ... 27 students enter a mathematics classroom, sit and quietly await the teacher's instructions.

. . . . . When all are ready, the teacher announces the day's plan of instruction. The student may or may not know the plan prior to coming to class ... but now it is clear.

. . . . . Class starts with a lecture by the teacher ... he uses the chalkboard or perhaps an overhead transparency projector. Perhaps a homework assignment is checked. Those students who have made errors now hopefully understand the reason for the error ... what about those students who have not made any errors?

. . . . . Following the lecture which has introduced a new math concept, the class receives an assignment in the textbook ... it is the state text--the main learning resource. The entire class is working on the same assignment ... some students breeze through ... some are having some difficulties ... the teacher is trying to help those students having difficulties.

. . . . . Before the end of the period, homework on the concept covered in class is assigned ... the bell rings ... the students leave.

. . . . . In time the students take a test. Some feel confident they have learned all the teacher will ask ... some are uncertain just what will be asked. Some students have studied hard, others very little. After the test student reaction is mixed ... some think the test was hard; others easy. The test results show some students received A's and B's ... some get D's and F's.

. . . . . The teacher attempts to help those students who have done poorly to understand the nature of their mistakes ... comments about how improvements may be effected are included.

. . . . . The next day a new math unit begins ... it is for all the students in the class ... it is likely the above process will repeat itself.

What Has Happened in This Classroom?

All the students are using the same materials and are

at the same place in the curriculum.   The teacher has spent
at least as much time with the group as with the individuals
in the group.   The instructional program is teacher-directed
consistent with the best interests of the majority.   The teacher has assumed the major responsibility for the learning
progress of the class.   If a student has not learned an important concept he must move ahead with the class.   The
brighter student is faced with the same problem of pacing
(and depth) ... he could move ahead faster than the class.

Depth of instruction and pacing are necessarily geared
to the majority.   The students are dependent upon the teacher
for their learning program--the teacher is the dominant force.
What happens if Johnny is not learning?   The teacher as a
professional does his best to reach Johnny, but how do you
reach 27 separate Johnnies in a class--each needing the attention and skill of the teacher?

A Glimpse at an ISI Classroom

. . . . .   The bell rings ... 54 students enter a double-sized
            classroom ... two teachers are in the room ...
            one is a math teacher, the other an English teach-
            er .. one part of the room is arranged as a math
            resource area, the other for English.   Students
            will spend the equivalent of two periods here.   The
            class is mixed in scholastic ability.
. . . . .   The students immediately go to work without teach-
            er direction.   Each student is working on his own
            learning contract.   His directions for learning are
            contained in his contract.   He knows the instruc-
            tional objective, the materials needed for learning,
            the way he will need to show he has accomplished
            the objective.   The teacher is available to assist
            him when needed.
. . . . .   Each student appears to be doing something differ-
            ent.   Some are working in English, others in
            math.   Some students are viewing a filmstrip,
            others are listening to tape recordings, others are
            working in programmed learning materials, some
            are writing, some are reading a chapter in a text
            or library book, some are working together, one
            child is seen helping another ... there is a low
            buzz of conversation as students discuss ... some
            students are moving from one resource area to
            another ... students are correcting some worksheets

from a key ... a teacher aide is correcting others.
Directed independent study is going on.

. . . . . A student completes his contract. He consults
with the teacher ... together they decide he is
ready for the post-test ... he passes the test ...
his progress is recorded on a wall chart ... he
is ready to begin the next contract. Some students
complete their contracts in a day or two, others
may take much longer ... the class is not affected
because the program is individualized, not group-
directed. A student fails the test ... in consultation with the child the teacher diagnoses the problem ... additional learning resources are prescribed ... perhaps the child learns better through
an auditory approach.

. . . . . The student schedules his own time for study in
each subject ... he knows his needs ... there is
no need to spend equal amounts of time on each
subject every day ... the scheduling can be varied.
The teacher monitors the scheduling process ...
he consults with the student to establish proper
time allotments if the student demonstrates a need
for greater supervision ... some students may require more guidance than others.

What Has Happened in This Classroom?

In the ISI classroom the student is assuming a substantial responsibility for his learning progress. The teacher
is his learning consultant. The teacher is sensitive to the
child's needs for inspiration and support. The teacher is
the child's resource for whatever guidance and assistance
may be needed. The student is practicing, under supervision,
the skills of learning that will have life-long implications.

If a child is absent, he has not missed a vital aspect
of the instructional program--he simply picks up where he
left off. All the resources for learning continue to be available to him because lecture by the teacher is minimal. At
the beginning of the school year, each child picks up where
he left off because the curriculum is organized without grade
level barriers. The attitude each child has toward himself
is improved because the program is built upon student successes rather than failures.

What Is the Rationale Behind the ISI Program?

The teaching-learning act: The ISI program is based
on the conviction that the purpose of any educational program
is to promote learning. The acts of teaching and the acts
of learning are not necessarily synonymous. Teaching is
supposed to be a means of causing learning; but it is possible
that what sometimes appears to be an act of quality teaching
does not produce the desired end product--optimum learning.

Some behavioral scientists say, "No person or machine ever teaches anyone anything at any time; ... learners
learn. Learning is an act of the learner, not of the teacher. "

The Borel ISI program focuses upon the learning act
rather than the teaching act. Inherent in this process is
determining the learning desired and providing the needed
learning materials and classroom climate to facilitate its
accomplishment.

The most significant aspect of this program is the
virtual elimination of the lecture technique as a means of
imparting knowledge to students. It has been said that when
teachers stop talking, students start learning. The lecture
is supplanted by resources that can be seen, read, heard,
or experienced--and repeated as often as the learner needs.
The lecture, or in ISI terms, teacher-led presentation, is
used only in very specific situations when needed learning
materials are not available.

The foundation upon which the ISI program is built is
the statement of learning objectives. These are written by
the teacher and seen by the student.

What are performance objectives and how do they differ from other educational objectives? Objectives, generally,
have not been stated in terms of what the learner is supposed
to do. Our objectives have usually been phrased as "The
student is to gain an understanding of the major concepts of
science"; or "The student is to increase his ability in critical
thinking. "

Performance objectives tell us how a student will show
us that he "understands. " They tell whether or not we have
succeeded in accomplishing the objective. Performance objectives tell what the student is supposed to do by behavior

that can be observed.

A sample performance objective: Given 20 sentences containing a variety of mistakes in capitalization, the student is able, with at least 90 per cent accuracy, to identify and rewrite correctly each word that has a mistake in capitalization.

Each performance objective has three elements: (1) the behavior desired, (2) the situation in which the behavior is to be observed, and (3) the extent to which the student should exhibit the behavior.

The development of sound performance objectives is probably the single most important and difficult assignment assumed by the ISI teacher.   Such objectives provide direction to both the learning experience and the appraisal of the effectiveness of a specific experience--the evaluation.

Studies have indicated that learning is more efficient when the student knows what he is expected to learn and how he will be expected to demonstrate the learning.   In the regular classroom setting, frequently a student may not know what he is expected to learn except vaguely.   This condition is best demonstrated by a student's attitude before a test. This often becomes an exercise in intuition.   What is the teacher's testing style?   Will there be an emphasis on memory or evaluation?   And when the test is over do too many A's mean the test was too easy or too many F's mean the test was too hard?   The well constructed performance objective and learning contract puts the student in a position of control--he knows what he must learn and how he will be expected to show it.

As the total curriculum is stated in a series of performance objectives, the teacher demonstrates his professional skill by knowing that some learnings are more basic than others; and some learnings are essential to future learnings. These factors may be written in to the routing of students through the series of contracts.

These objectives recognize that learnings may be placed in three categories: (1) MUST KNOW--learning acts that are essential; (2) OUGHT TO KNOW--learnings that most students should accomplish; and (3) NICE TO KNOW--non-essential learnings that represent deeper and finer understandings.   The placing of learnings into these categories is difficult. There may be many who say it is unnecessary, impossible, and perhaps undesirable. But I believe it is a necessary formidable act if we are to indeed personalize our instructional program.

The pre- and post-evaluation processes that are normally associated with a unit of instruction must give thoughtful attention to these categories of learning. We must avoid the tendency to test NICE TO KNOW and OUGHT TO KNOW items to the virtual exclusion of MUST KNOW items.

We court eventual disaster if we permit a youngster to progress to increasingly more difficult concepts when he has not demonstrated that he has mastered a prerequisite MUST KNOW concept.

This is a significant point, especially for the less successful student. The group-directed teaching situation is poorly equipped to deal with such learning omissions which eventually become tragic gaps in needed learning skills. The class must move on; it cannot wait for one or two students. Before long such students find themselves faced with the impossibility of being expected to learn something for which they are not ready. Such students either stop trying to learn or else make desperate attempts at memorizing materials that have little meaning for them.

Since the ISI program is individualized and non-graded, the student may indeed progress at his own pace and utilize the techniques of learning that are best suited to him. Those who need more time, get more time. Attention to accommodating each youngster's own style of learning is possible.

Until very recently, the chief sources of learning in schools were the teacher and the textbook. There were other sources but they were spotty and incidental and not integrated into the mainstream of the instructional program. So long as the teacher and textbook play dominant roles as sources of knowledge, individualization is impractical and faces eventual failure. A teacher cannot continue to function in the same way and accomplish substantial individualization.

Advances in technology have made available to the instructional process a wide variety of instructional materials beyond the limitations of the traditional textbooks. Without an ample and varied battery of instructional tools, an individualized program will founder. The media utilized in the

ISI program may include: programmed learning materials,
filmstrips, filmloops, tape recorders, synchronized slides
and tape recordings, records, films, and teacher-prepared
materials.

The teacher continues to be the single most important
factor in the excellence of an instructional program.   Needed
teacher time and energy for individualization cannot easily be
realized as long as the teacher accepts the role of the chief
imparter of knowledge.   Only when his own expertise is
needed should he use brief teacher-led presentations to im-
part knowledge.

The ISI teacher role is different.   Relieved of the
daily demands to prepare, adapt, and present information,
he serves in a role only he and no one else can serve:
director of learning.

. . . . . he prescribes the course of study
. . . . . he monitors each child's progress
. . . . . he diagnoses learning problems
. . . . . he prescribes possible alternative learning mate-
           rials and activities
. . . . . he evaluates each child's progress in achieving
           behavioral objectives

Thus the teacher may be seen working with individuals.   He counsels a child regarding the scheduling of time.
He uses a firm hand or a gentle touch depending upon each
child's needs.   The talents of the teacher may be better
utilized with greater efficiency in a one-to-one relationship
than in a one-to-ten or one-to-30 relationship.   One minute
spent with a child at his moment of need may be more worthwhile than 15 minutes spent with the entire class.

Children need group activity as much as they need
directed independent study.   The teacher who is sensitive to
this need provides opportunities for it.   Group activity will
occur as small groups engage in discussion where children
actively listen and participate.   Group discussions on sections of the objectives permit student-to-student learning
which is very important.

ISI is not tied to team teaching or flexible scheduling
although they are often associated with it.   The ISI program
may be utilized in a self-contained classroom and within the
equal time periods normally found in a departmentalized

program.   In several of our ISI programs, the element of
team teaching and flexible scheduling will be used.   In these
situations we expect to use several adjacent open space
classrooms to which students are assigned for perhaps a
three-period block of time.

Within the open space, learning resource areas for
each subject are established.   The student, guided by his
own learning contract, schedules his own time for each subject.   Team teaching techniques are therefore utilized as
teachers plan together for the learning progress of all their
students.   Flexible scheduling is accomplished in the sense
that the student need not stop learning math at the end of a
fixed period; he may continue working with his math contract
for 100 minutes today, while tomorrow it may only be 20
minutes.   This is the ultimate in effective flexibility of time.
There is no pre-determined number of time modules allotted
to subjects throughout the week; there is no "free time"
within the time block.   The student works under the general
supervision of his teacher (directed independent study).

Can Students Handle Responsibility?

The fact that students are expected to assume substantial responsibility for their learning progress is one of
the major strengths and objectives of the ISI program.   But
the teacher shares this responsibility--the student is not left
alone without guidance and supervision.

Because the ISI program is personalized, it permits
a variety of approaches to be used by the teacher in monitoring the progress of each child and guiding his progress effectively.   Many students can assume responsibilities with
minimum teacher supervision.   Others cannot.   The teacher
is free to use whatever technique best suits the needs and
characteristics of each student.   The ultimate goal is to
develop in students the attitudes and techniques to build responsibility, motivation and achievement.

Toward this end, the teacher and child will probably
set goals that are mutually agreed upon.   The student must
share in this process if the ultimate goal is ever to be accomplished.   The "force-feed" approach that satisfies the
moment, may, for some children, be the poorest technique.

What about Marks, Grouping, and Homework Practices?

One of the advantages of the ISI program is the minimum need for ability grouping or tracking students in classes.
This practice is quite often necessary in group-directed
classes.   Why?   While this may appear to be an arrangement designed to improve learning, problems quite often appear in those classes whose curriculum may be described
as remedial.   The learning potentialities of a youngster are
quite often affected by his own opinion of his potential.   When
a youngster finds himself in such a class, even though the
curriculum may be appropriate for him, he tends to feel
stigmatized, develops a defeatist attitude and behaves accordingly.   The student who is in a class with all kinds of
ability levels avoids the label of the "low" class and quite
often raises his expectations and begins to learn and behave
accordingly.

It is no secret that youngsters of this age are strongly
influenced by their classmates.   If a student is part of an
environment that is characterized by high expectation, strong
motivations, and good learning habits he seeks to assume
these values even if he did not previously display them.

The ISI program can be organized with a heterogeneous grouping only because the instructional program is individualized.   Grouping need not be as high or low classes.
The learning needs of each student regardless of his level of
achievement can be met.

Homework is handled on an individualized basis.   Some
students may not have homework to any significant extent,
while others will.

In conclusion, the potential advantages for all sorts of
students--high, average, low--are limitless in a personalized
instructional program.   Our task is now to re-tool the instructional process to implement the program.   The process
may be difficult and sometimes painful, but it must begin and
it must begin now.

AN INTEGRATED HIGH SCHOOL PROGRAM
OF INDIVIDUALIZED INSTRUCTION

J. Clair Morris

A. Definition of Terms

INDIVIDUALIZED INSTRUCTION: a process designed
to educate each student at a rate and depth commensurate
with his physical, social, emotional, and intellectual growth.
It does not mean, however, that a teacher works with only
one student at a time. It does mean that a student works on
content for which he is educationally ready and in a manner
which is most efficient for him.

Individualized instruction is generic in nature and,
therefore, includes a variety of methods for its achievement.
For example, phasing, flexible scheduling, nonogradedness,
continuous progress, small group instruction, independent
study, team teaching, and vocational programs are all methods and organizational structures which enable individualized
instruction to be present. In this report, the terms individualized and instruction are used generically and include an
educational program comprised of the sub-programs just
cited in this paragraph.

CONVENTIONAL METHOD: a teaching-learning process in which an instructor teaches approximately 35 students
via a group method. In this method all students are expected
to proceed through a prescribed content at the same speed
and depth. This is usually accomplished with a set of identical textbooks.

INDEPENDENT STUDY: these terms have reference
to a student study process in which individual students study
in semi-isolation rather than in small groups or in large
groups. To enhance this process, time is provided via unscheduled time within the school day, via released time from
regularly scheduled classes which enables students to move
physically to independent study stations, or via time which
is provided in the regular classroom for study purposes.

SMALL GROUP INSTRUCTION: this organizational
structure enables students with similar achievements,

interests, or needs to come together and learn via frequent interactions with each other. Groups usually include from two to 15 students.

PHASING: a system of grouping students according to different achievement levels based on both subjective and objective data. Individual differences, under this grouping plan, are intended to be partially met by providing courses in the curriculum of varying hardness levels for students of different achievement levels. Mobility among phases is encouraged for students who illustrate a need for such changes.

NONGRADEDNESS: a nongraded program is one in which the organizational structure has been altered to enable students to register for classes for which they are educationally ready. This program can be contrasted with a graded school in which specific classes are prescribed for each grade level regardless of the interests, desires, or educational readiness of the students. Nongraded programs enable students to register for any class for which they are ready during any year that they attend school.

TEAM TEACHING: a means of teaching wherein two or more teachers cooperatively work with the same number of students as they would teach were they working individually rather than in teams. Such a cooperative plan makes possible flexibility in the grouping of students. The students can be in large groups for efficiency of time while a lecture is being given or may be broken into small groups for small group instruction or seminars. Independent study is also possible under this organizational structure. Teacher strengths are utilized for a larger number of students in team teaching than is the case in a conventional program.

FLEXIBLE SCHEDULING: in this study, two types of flexible schedules are used. One is a hand-made modular schedule which provides for different time lengths for classes in the curriculum. The schedule is the same each day; however, teachers having classes several modules in length have a daily option relative to the amount of time which will be used each day. The second type of flexible schedule used in this study refers to one which contains class periods of variable length which rotate daily. The schedule makes a complete rotation in one week, thus giving each teacher and each class a turn at each different class length each week. At the time of this study, other types of flexible schedules were in operation in other parts of the United States.

CONTINUOUS PROGRESS: these terms refer to an organizational plan which enables students to progress through the curriculum at rates of speed and at content depths commensurate with their interests and abilities. This approach contrasts with the conventional method which moves students through the curriculum as a total group.

B.   The Educational Program for 1962-63

The educational program for the 1962-63 school year was conventional or traditional in nature. By conventional it is meant that it was characterized in general by individual teachers working by themselves with groups of 30 to 35 students. Team teaching, phasing, and continuous progress programs were not present. The students usually worked from common textbooks and progressed through the curriculum at the same rate of speed. Independent study and small group instruction may have taken place in some individual classrooms as has always been the case. The rotating schedule provided time for independent study four periods per week. Some nongrading was present as had been the case in Cedar High for many years prior to the 1962-63 school year. An analysis of the schedule for 1962-63 showed that 50 per cent of the courses were graded and 50 per cent nongraded.

There was a rotating schedule which provided an element of flexibility. This schedule was also present during the 1961-62 school year. The schedule provided class rotation; therefore, the same classes were not held daily at so-called "bad times" of the day such as the last period in the day or just following lunch. The schedule included four 55-minute periods, one 35-minute period, and one 70-minute period. It provided for individual counseling, group counseling, class meetings, independent study, and assemblies. The last, or sixth, period did not rotate because of athletics. All students ate lunch at the same time.

C.   The Educational Program for 1964-65

The 1964-65 school year was the first full year that the new building was occupied. The new building was constructed on a 39-acre site. The building contained 222,680 square feet, or 5.1 acres of improved and unimproved floor space. The total cost of the building was $3,052,181.30.

The building was two and one-half years in the planning stage and two years in the construction stage. The basic philosophy of individualized instruction dominated the educational specifications. The educational specifications were cooperatively planned by administrators, teachers, students, parents, and personnel from the School Plant Planning Laboratory at Stanford University. Cedar High School had a $5,000 grant from the School Plant Planning Laboratory to assist in preparation of the educational specifications.

The building was designed for flexibility so it would house a conventional curriculum program or a highly innovative program featuring organizational changes such as team teaching, large group instruction, seminars, and independent study. The building, when completed, contained a 982-seat auditorium and two large classrooms which would seat 150 and 93 students respectively. Included also were 42 regular-sized classrooms, 14 small group seminar rooms, and three large rooms adjacent to the instructional materials center which contained 340 individual study carrels. Obviously, the structure would accommodate large group instruction, independent study, and conventional instruction in classrooms designed for one teacher and 35 students.

Upon completion of the facility, the major task was one of implementing an educational program based on meeting the individual differences of students. The 1964-65 school year contained the first major program changes based on individualized instruction. Most of the changes started out on a miniature basis in a single department and then spread to other departments in the years which followed. The first year of the experimental program included programs in team teaching, independent study, small group instruction, phasing, nongraded classes, continuous progress programs, and a rotating schedule which contained six periods. There was one vocational services class for girls and several industrial arts classes for boys.

TEAM TEACHING: a strategy present only in the English department.
INDEPENDENT STUDY: as a teaching means, present in the English department, which included 515 of the 555 students in the school.
SMALL GROUP INSTRUCTION: presently only in the English department.
PHASING: in the 1964-65 school year, present for the first time in the English and biological science departments.

NONGRADED CLASSES: classes nongraded for the first time were all English language arts classes.

CONTINUOUS PROGRESS PROGRAMS: found in the English department as a planned part of the educational strategy. Students were also in continuous progress programs in Spanish and German which were greatly assisted by the foreign language laboratory and the many accompanying magnetic tapes. Other continuous progress programs were present in art, crafts, woods, metals, and clothing which included students who worked on individual projects. Programs in these areas were not unique because they have probably always been handled on an individual project basis. The unique programs were in English, German, and Spanish.

ROTATING SCHEDULE: designed and implemented to provide an element of flexibility in class length. The schedule included six periods, of which five were 50 minutes in length and one was 75 minutes in length. On Monday, for example, the first period classes came first in the day and were 75 minutes in length. On Tuesday, the second period classes came first in the day and were 75 minutes in length. The third period remained constant each day because it was back-to-back with the lunch period. One-half of the students went to lunch first and then went to third period; the other half went to third period first and then to lunch.

The rotating schedule enabled each class period except the third to be 75 minutes in length once per week. Teachers could then plan to have films, short trips, etc. on days when it was their turn to have the long period. There was also a benefit, according to teachers, in having each class held at different times during the day on a weekly basis. For example, no class was always last each day or just after lunch. Students and teachers were able to adjust to the changing schedule each day without difficulty. The schedule made a complete rotation each week.

D.   The Educational Program for 1965-66

Innovative programs present during the 1965-66 school year included team teaching, independent study, small group instruction, phasing, nongraded classes, continuous progress programs, a seven-period day, and added vocational programs. Of special interest is the fact that the English department abandoned the experimental program and went back to a more conventional approach to teaching.

TEAM TEACHING: present in the biological science
department and in one section of a vocationally oriented program for low-achieving tenth grade students.   There was
cooperative team planning in the English department but no
actual team teaching.

INDEPENDENT STUDY: this was enhanced by a seven-
period day.   Students were required to take six classes and
could take seven classes if they desired.   If students did not
register for the first class in the morning, they could stay
home until second period.   It was also possible for students
to go home at the end of the sixth period provided they were
registered for the first six classes.   If they had classes the
first and seventh periods, it was possible for them to have
an unscheduled period during the school day which could be
used for independent study.   Approximately 175 students had
unscheduled periods for independent study.   Carrels were
provided for this study time.   Independent study time was a
part of the teaching strategy in most classrooms which
amounted to about 10 per cent of the regular class time as
estimated by teachers.

SMALL GROUP INSTRUCTION: very little took place
in the 1965-66 school year.   The only small group instruction
present took place in classrooms in which a single teacher
worked with a group of approximately 30 students.   This
situation was not unusual enough to define as an innovation.

PHASING: present in biology, in English, and for the
first time in American problems.

NONGRADED CLASSES: approximately 70 per cent of
the courses in the curriculum were nongraded during the
1965-66 school year.   Courses in biology, American history,
and American problems were still graded for ease in gathering attendance statistics for state attendance reports.   Pre-
requisite courses which were preparatory for advanced
courses were bypassed if students showed themselves to be
educationally ready for the advanced courses.

CONTINUOUS PROGRESS PROGRAMS: available in
geometry, remedial reading, clothing I and II, foods, auto II,
drafting, typewriting, German, Spanish, shorthand, office
practice, art, crafts, woods, and metals.   Most academic
classes were handled on a total group basis.   This was due
to a lack of continuous progress designed materials.   The
Edison remote controlled dictaphone system was purchased

during this year which made it possible for shorthand students to receive at their desks four different dictation speeds. This system helped to care for individual student differences in shorthand.

English department reverts back to conventional approach: as was noted in the description of the 1964-65 school year, the English department was actively involved in team teaching, small group seminars, and continuous progress programs via independent study. It would be unfair to say that the program was completely unsuccessful. Yet, so many problems were encountered that the English staff and administration decided it was best to change the program for the 1965-66 school year.

SCHEDULING: the schedule for 1965-66 contained no flexibility. It was a seven-period day, each period being 50 minutes in length.

E.   The Educational Program for 1966-67

The 1966-67 school year included programs in team teaching, independent study, small group instruction, phasing, nongraded courses, continuous progress programs, an expanded vocational education program, and an eight-period day.

TEAM TEACHING: present in the biology, English, and the social science departments.

INDEPENDENT STUDY: present as part of the teaching strategy in most classrooms. Independent study also took place in student study carrels during each of the eight periods each day. Of 575 students, 240 had at least one period during the day which was unscheduled and which could be used for independent study.

SMALL GROUP INSTRUCTION: present in English classes, reading classes, and social science classes.

PHASING: This type of grouping, designed to complement individual differences in students, was present in the biology department, in general mathematics classes, in English classes, and in social science classes. The general mathematics classes were divided into phases I, II, and III with phase III being the most sophisticated in content. There was one section of phase I students, two sections of phase

II, and one section of phase III.   The total number of students involved was 111 or about 19 per cent of the total student body.

NONGRADED PROGRAMS: during the 1966-67 school year, about 65 per cent of the classes offered in the curriculum were nongraded.

CONTINUOUS PROGRESS PROGRAMS: included were classes in geometry involving 44 students; algebra, 20; reading, 24; American problems, 15; American history, 31; woods II, 14; metals, 16; services math, 20; drafting, 35; typewriting, 48; shorthand, 39; clothing, 65; foods, 27; special education for the mentally handicapped, 12; Spanish, 20; and crafts, with 77 students.

F.   The Educational Program for 1967-68

The educational program for the 1967-68 school year included programs in team teaching, independent study, small group instruction, phasing, nongradedness, continuous progress programs, and modular-type flexible scheduling.   Vocational education programs were greatly increased.

TEAM TEACHING: included in the departments which had enough teachers and students available to make team teaching organizationally possible.   Areas with these conditions present included the biology, English, and social science departments.

INDEPENDENT STUDY: during the 1967-68 school year, independent study took place within regularly scheduled classes and during unscheduled time modules.   Independent study within classes was not the same each day relative to quantity; however, teachers estimated that about 15 per cent of the students' time during class was used for independent study.   Unscheduled time which resulted from students not being registered for classes during some of the modules each day averaged out to be 17 per cent of the average student's school day.   A school day was defined to be from 8:40 a.m. through 3:30 p.m.

To determine how unscheduled time was being used, a survey was conducted on January 4, 1968.   The survey included 680 of the 687 students enrolled in school.   Each student was instructed to allow 40 minutes for lunch and

designate that time as scheduled time.   Student helpers were
posted at locations where students were encouraged to go during unscheduled time.   Student helpers were also posted at
locations such as the school parking lot and the bowling alley
downtown to observe students in those areas.   The library
area was the most popular, being visited by 596 students.
It was also observed that 28 students visited the community
bowling alley, an area in which they were not supposed to
be present.   The students had 22 locations where they could
legally go on unscheduled time provided there was available
space and provided they had permission from the teacher who
had supervision over the area.   The survey had the limitations of being conducted only one day and, therefore, may or
may not reflect what happened on the other 179 days in which
school was in session.

PHASING: a system of homogeneous grouping via
achievement was implemented in the biology, English language arts, general mathematics, and social science departments.   General mathematics classes included phases I, II,
and III.   Seventy students were in phased general mathematics classes or about 12 per cent of the total student body.
Students in biology, general mathematics, American history,
and American problems were phased according to past
achievement in classes and on standardized tests.

CONTINUOUS PROGRESS PROGRAMS: implemented
in English, American history, American problems, Spanish,
advanced auto, advanced metals, trigonometry, shorthand,
geometry, algebra II, typewriting, pre-technical science,
art, crafts, foods, clothing, drafting, electronics, woods I,
algebra I, and marketing.   All the students of the school
were included in continuous progress programs in one or
more departments of the school.   Teachers issued credit to
students as courses were completed.   The credit incentive
was based on achievement rather than strictly on time.

A major deterrent to an increase in continuous progress programs was still a lack of curricular materials sufficiently self-explanatory to enable students to progress at
their own speeds without continuous teacher assistance.
These study guides were just not available in quantity for
teacher use.

MODULAR TYPE FLEXIBLE SCHEDULING: the only
year this scheduling was in use was 1967-68.   The objective
in creating the schedule was to enable different classes to be

held different lengths of time, depending on the needs of the
different types of classes.   The modular schedule achieved
the objective of variability relative to the time blocks.

Each module was 20 minutes in length, except the attendance and counseling module which was 15 minutes in
length and the first module which was ten minutes in length.
The schedule included 36 modules, six of which were in the
evening.   The first module began at 7:30 a. m. and the last
module ended at 9:15 p. m.   Students started school at different times in the morning, went to lunch at different hours,
started at different times in the afternoon, and went home at
different times at the close of the school day.   Some high
school students had a night class.

Since most modules were 20 minutes in length, classes
could be any multiple of 20 minutes in duration.   One class,
clothing theory, was one module in length.   Many classes in
language arts, social studies, vocal music, and typewriting
were held for two modules.   Physical education, band, or-
chestra, cosmetology, a cappella choir, phase I biology, and
home economics classes were held for three modules.
Classes needing four modules were phase II and III biology,
chemistry, and physics in which longer blocks of time were
needed for laboratory activities.   Advanced vocational classes
such as electronics, drafting, auto, carpentry, metals, office
procedures, marketing, sewing, nursing, and fry cooking
were five modules in length.   One class, office machines
repair, was six modules in length two evenings per week.
Community sports, a nine module class, was held two after-
noons per week.   Most classes ranged from two to five modules in length.

Achievement

Significant differences were not present during the
five-year period in the areas of mathematics, science, social studies, or writing, between the individualized and conventional methods of instruction.   The null hypothesis that
no significant differences would exist was not rejected.   Relative to the four areas involved, there is no logical reason
to say that one method was better than the other method.
Males achieved significantly higher scores in the areas of
mathematics, science, and social studies than did females.
The differences were especially evident among the above
average ability students.

Students scored significantly higher in critical thinking
skills in the convention year of 1962-63 than they did in the
years of 1965-66 and 1967-68, during which the individualized
method was in operation. The 1962-63 group did not score
significantly higher than did the 1966-67 and 1964-65 groups;
however, the conventional group did, in each case, have
higher mean average scores than did any of the individualized
groups. The null hypothesis was rejected in favor of the
conventional group. Relative to critical thinking, the conventional approach was superior to the individualized approach.

The 1962-63, or the conventional group, scored significantly higher in study habits and attitudes at the .05 level
than did either of the individualized groups which completed
the questionnaire in 1965-66 and 1966-67. The null hypothesis was rejected in favor of the conventional group.

The conventional group of 1962-63 scored significantly
higher in library skills at the .01 level than did the individualized group of 1967-68. The null hypothesis was rejected
in favor of the conventional approach which proved to be
superior to the individualized method relative to library
skills.

Educational aspirations were determined by an eight-
item questionnaire. The null hypothesis that significant differences would not exist between the individualized and conventional methods was not rejected by a chi square analysis
of each of the eight questions. It could not be said that
either method was superior to the other relative to educational aspirations.

There were significant differences in the proportion of
student dropouts during the five years of the study. The difference was significant at the .01 level. Chi square comparison of the conventional group with the four individualized
groups yielded a significant difference at the .001 level in
favor of the individualized approach. The individualized approach was superior to the conventional approach relative to
the proportion of dropouts.

Overall Summary

Results were not significantly different between the two
methods in the six areas of mathematics, science, social

studies, writing, educational aspirations, and sociometric relationships.   Results in three other areas, reading, listening, and attitudes toward education showed vacillation when different years were compared.   Three of the comparisons on reading were not significant and one comparison was significant in favor of the individualized approach.   In listening, three comparisons were not significant and one comparison significantly favored the conventional approach.   Two comparisons on attitude toward education were not significant and one comparison was significant in favor of the conventional approach.   The three areas of critical thinking, study habits and attitudes, and library skills significantly favored the conventional method.   The one area of dropouts significantly favored the individualized approach.

THE UNIVERSITY OF PITTSBURGH MODEL OF TEACHER TRAINING FOR THE INDIVIDUALIZATION OF INSTRUCTION

Charles J. Gorman

Individualized instruction has been sought by many teachers during the short history of American education. Through the years, volumes have been written on this concept and glib speakers have urged the implementation of an instructional program geared to each learner.   Unfortunately, very few examples of genuine individualization can be found today in the schools of our country.   The University of Pittsburgh model of teacher training has been prepared with individualized instruction as the central theme.   It is hoped that this training model will make a significant contribution to the implementation of individualized instruction.

A general definition of individualization, adopted in the model, is as follows: Individualized Instruction consists of planning and conducting, with each pupil, programs of study and day to day lessons that are tailor-made to suit his learning requirements and his characteristics as a learner. Thus, by definition, the individualized instruction which has been conceived in this model is marked first by planning and then by implementing the plan.

Model Features

Four structural features dominate this model for teacher training. In two of these features, flexibility and self-development, personal needs have been recognized. With the other features, mastery and efficiency, professional qualifications were acknowledged.

Flexibility was viewed as an essential feature in any endeavor which honors individualization. This attribute is evident in the model as such procedures as Alternate Learning Routes were incorporated in various learning modules. In this manner, different rates and styles of learning were accommodated.

Self-development was featured in several phases of the program. The selection process incorporates it through assessment of potential candidates. By this process, training can be adjusted for each student. Extensive group process experiences also focus on this area. Through such techniques, students learn how to help others in a group or team setting.

The Pittsburgh Model also characterized the concept of mastery. Trainees will be expected to demonstrate that learning goals have been met and movement through the program will be predicated on the evidence of mastery of specified learning goals. However, rigid standards of performance for all trainees will not be used.

Efficiency is the final feature of this teacher training plan. This trait is related to the notion of flexibility. Efficiency is a prime feature for it refers to the practice of adjusting to individual knowledge, learning style, and interests. In this way, undue delays and unnecessary repetitions are avoided.

Requirements

The requirements of this model are classified under five interrelated categories. As a whole, they form a network consisting of cognitive input, affective experiences, and field participation. The specific requirements are (1) academic education, (2) professional education, (3) teacher competencies, (4) a guidance component, and (5) a clinical setting.

1.   Academic education refers to the liberal arts specifications.   Included in this domain are communications, humanities, social sciences, and natural sciences.   While the content of this area is not noticeably different from the past, it is proposed in the model to change the manner of teaching the liberal arts specifications.

2.   Professional education includes the study of learning theories, child development, psychology and all other areas related to teaching children.   In one sense, this requirement could be labeled the "knowledge base" of teaching. Education presently lacks such a base.   This condition can be improved upon by the establishment of a systematic feedback process which monitors the training program.   The Pittsburgh Model includes the strategy to initiate this process.

3.   The third requirement refers to teacher competencies, which are described in the form of behavioral outcomes.   The nine categories include (1) specifying learning goals, (2) assessing pupil achievement of learning goals, (3) diagnosing learner characteristics, (4) planning long-term and short-term learning programs with pupils, (5) guiding pupils with their learning tasks, (6) directing off-task pupil behavior, (7) evaluating the learner, (8) employing teamwork with colleagues, and (9) enhancing self-development.   In designating this list of behaviors, the model builders acknowledged the open-ended nature of each category.   Research and experience will enable further clarification of the role of the teacher for individualized instruction.

4.   The guidance requirement includes group process experiences, individual counseling, and group directing.   Self-realization, self-development, and self-evaluation are major goals of this section.

5.   An adequate clinical setting has been described in the model as one which grows out of agreement by the university, school district, professional organization, and governmental agencies.   A new form of cooperation has been proposed around the central purpose of this model--the individualization of instruction.

Student Progress Through the Model

In general, this model follows the basic procedures of

most instructional models, i. e. , trainees are provided experiences of an instructional nature in order to change their behavior as indicated by the specific goals and objectives of a program.   The academic education requirement is the dominant theme of the first two years of training.   Toward the end of the second year, the trainee indicates an interest in the teaching profession.   At this time, a thorough admission process is initiated which includes experiences in the clinical setting for each candidate.   The final two years of pre-service education includes a focus on professional education, teacher competencies, and self-development through the guidance requirement.   Most of these experiences occur in the clinical setting.   The trainee experiences several roles during the final two years such as assistant teacher, student teacher, and intern.

As the trainee participates in the clinical setting, the dominant features of this model--flexibility, self-development, mastery, and efficiency--are manifested throughout the process.   The length of time of student teaching will be adjusted in accordance to the needs of the trainee.   Long-term group process experiences will be provided to avoid a superficial sensitivity to self.   Evidence of specific competencies will be sought and provisions will be made for the trainees style of learning and operational level.

Summary

The University of Pittsburgh model of teacher training for individualized instruction is a general plan.   Elaborate units or extensive instructional modules have not been prepared because the model builders view the development of such instructional materials as the necessary experience of all faculties interested in the individualization of instruction.

One vital agreement reached by the team which built this model was that trainees must witness individualized instruction throughout the pre-service experiences.   In this way, the concept of individualization likely will be internalized.   Thus, it is assumed that graduates of this experience will make a significant contribution to the implementation of procedures leading to individualized instruction in the schools of America.

IV. INDIVIDUALIZING MATHEMATICS INSTRUCTION:
   1. SOME EXAMPLES

        In the six selections found in this chapter we have
an opportunity to view individualized instruction in elementary
school mathematics from a variety of viewpoints: a principal
and a college faculty member report on one program; a college professor describes and evaluates a program in a campus school; a sixth-grade teacher tells of his experiences in
using an individualized approach; and two writers of doctoral
theses describe the implementation and effects of the same
plan for individualizing mathematics instruction.   These selections also span the decade of the sixties; the earliest article was written in 1960 and the latest is taken from a 1969
publication.   In addition, all the elementary grades are involved in one or more of the descriptions of individualized
programs.

        In all but one of the articles in this chapter a situa-
tion is described in which there was an ongoing program of
cooperation between the school and a university.   As one who
has taught in the elementary school and who has been a member-
ber of the education department of a college for the past
several decades, I cannot stress too strongly my feeling that
there is much to be gained by fostering such relationships.
Not only do the children and the teachers of the school have
an opportunity to profit from such an arrangement, but there
is also a great gain in the effectiveness of the college personnel assigned to the implementation of such an arrangement which, of course, results in better training of those
who are being prepared to become the teachers of tomorrow.

Abraham Kaplan and Marilyn Jasik

        The first two articles in this chapter are descriptions
of the operation of an individualized instruction program in a
public school in Queens, New York which was taking part as a

School University Teacher Education Center (SUTEC) in a co-operative project between Queens College of the City University and the Board of Education of New York City.

The first selection, by Abraham Kaplan, then principal of P. S. 79, Queens, is taken from his article that appeared in High Points, a publication of the New York City Board of Education. I think that there are several points in his discussion that are particularly worth noting. For one thing, he paints a drastically contrasting picture of group and individualized procedures which deserves the reader's careful examination and analysis. And for another, he makes the point that at first only ten teachers took part in the individualized procedures introduced by the Queens SUTEC team but that subsequently it was adopted by all classes. This is one possible answer to the question that is so frequently put in one form or another: Where do you start an individualized program? In the bright classes? In classes taught by experienced teachers? In classrooms where the teachers volunteer to take part? Or do you start with the school as a whole?

There certainly does not exist any one pat answer to these questions. As I have tried to emphasize previously, individualization is never a program with fixed, unalterable sets of rules and regulations. The basic point of the philosophy of individualization is flexibility for all concerned. This, of course, includes the school principal, who will make a decision based on factors that are unique to the particular situation that he faces. I have known of a number of schools where new programs of any kind are introduced in the classrooms of those teachers who volunteer. I must say that it has been my observation that this does not invariably result in a school's wide adoption of the innovation within a short period of time. On the other hand, as I pointed out in my discussion of the Kramer article in the preceding chapter, it is a risky procedure to impose a program on unwilling teachers.

I also will comment briefly on the notion that if certain classes or certain teachers are to be selected to initiate the implementation of an innovative program, either the classes or the teachers need to be selected on the basis of any one or more particular pre-determined criteria. Starting, for example, with the "bright" classes may lead to the judgment by other teachers that the program may be effective only in such classes, and believing this, these other teachers may become even more resistant to the adoption of the

innovation than they already were. Similarly, I doubt that
teachers should be chosen on the basis of level of experience. If a criterion other than willingness as evidenced by
volunteering is to be used, the decision should be based on
the best administrative judgment of the administrator or
supervisor rather than on any predetermined criterion.

Marilyn Jasik, who also describes the program about
which Mr. Kaplan wrote, was with the Queens College Department of Education for five years before moving to Israel
where she now is based. She writes me:

"My concern with individualized instruction, better
called personalized instruction--because after all it is the
person first and the instruction second as I see it--stems
from my first love in education, the early childhood years.
We would never think of lockstepping a group of three-year-
olds because we are so conscious that their learning patterns
are terribly individual. Somehow, these children grow a bit,
we get them into larger classes and bigger buildings, and in
our American fashion, we search for systems that serve the
organization but, in far too many cases, illserve the individual.

"I believe this strongly--from all my observations as
a classroom teacher, as a college instructor and supervisor,
as a student myself, and as a mother. There is no single,
all-inclusive learning potion that works for all children. And
awareness of this fact calls for a reassessment of our teaching techniques. "

It is not at all surprising to me that one who writes
with such an obviously deep conviction should be sensitive
enough to select the incidents described in her passage to
emphasize the benefits of individualization in the early school
grades.

Helen Redbird

In this chapter's third selection, Professor Helen Redbird of the Oregon College of Education at Monmouth, Oregon, describes a two-year individualized mathematics project
conducted in the fourth grade of the Campus School of that
college.

No one can, I think, seriously question the merits of

and the sincerity in the effort in the program described in this very fine article. In evaluating this report there are, however, certain items that must be considered:

(1) The concept of the Campus School is one that has great appeal to trainers of prospective teachers. Here is an opportunity to show how it should be done in a situation where experimental procedures are accepted without question and where the college has jurisdiction over the teachers and over their in-school activities. While this may be an ideal way of showing prospective teachers how it could be, there are many who, like myself, feel that this ill prepares a young student to become a teacher in a school where the climate is not in any respect similar, where innovation is regarded as undesirable, where advice by college teachers is regarded as unwarranted interference, and where the young teacher fresh out of college is restricted by rules, by customs, and by social pressure from departing to any degree whatsoever from the manner in which classes of that school have operated since the memory of man runneth not to the contrary.

The argument is advanced that the college student is far better off to observe, tutor, and do student teaching in the second type of school where college personnel supervising the students' activities could at least show some of the ways in which the young teacher can serve as a change-agent even in an almost hopeless situation such as I have tried to describe in the previous paragraph.

(2) Of course every teacher dreams of what could be accomplished under ideal conditions with a crash program where every necessary resource is brought to bear on an educational situation. In this passage Professor Redbird depicts such a saturation process in action. The question is what can we legitimately conclude from reading the remarkable results obtained. We know that in the real world such saturating procedures are impossible. Does this passage tell us that there is or that there is not hope in that real world for attaining like results?

These objections may occur to the readers as they have to me but I hope that you will also share with me a sense of excitement in reading an account of a study that shows that results of the type described here can be obtained. To me this says that if they can be obtained in a way that is not realistically available for my use, then there certainly must be hope that they can also be attained in other ways.

Certainly individualization of instruction is not practicable
only under the ideal conditions described here.   I ask myself
if it is possible that the real crucial factor here could have
been individualization rather than the saturated resources
that were brought to bear.   It seems worth thinking about.

Walter L. Whitaker

The following article was written by a sixth
grade teacher who describes the procedures he used to individualize the teaching of mathematics in his classroom and
tells why he feels that that which he did was sound.   It is
good to read a report by a practitioner who is not isolated
from reality but is actually there (where the action is!).   It
is good to read his report that he feels that individualization
has a place in a real life situation.

Mr. Whitaker is now a coordinator in the Culver City
Unified School District in California where the work reported
in this article was also done.   He writes me:

"It might be of interest to you to note that in the
early sixties, when I was carrying off the individualized program in a sixth grade class, these articles and my experimentation were considered heresy by my colleagues and some
of the local administrators.   It is of interest to me that only
a few years later, many of the ideas and things which I tried
are becoming fashionable and are being incorporated into
various attempts at improving the curriculum. "

Elaine V. Bartel and K. Allen Neufeld

The last two passages of this chapter are excerpted
from two theses, sponsored by Professor M. Vere De Vault
at the University of Wisconsin in 1965 and 1967.

The author of the earlier thesis is now on the staff of
the University of Wisconsin in Milwaukee where she serves
as director of the Intern Teaching Program.   She writes me:

"I have served as a consultant to elementary schools
in Wisconsin as they move toward programs of individualization.   I have conducted undergraduate and graduate workshops
and seminars in individualization with emphasis on programs
of reading and mathematics, but including all curricular

areas. "

      Professor Bartel, in the excerpt from her thesis in-
cluded here, describes the individualization of mathematics
instruction at the fourth-grade level in several public schools.
The plan which is described was known as the Individualized
Mathematics Curriculum Project (IMCP) and was developed
at the University of Wisconsin. Our particular interest in
this passage lies in the fact that while we have been skirting
the issue of self-selection in discussing the previous material we now come upon an account of the implementation of
a program whose very essence is the principle of self-selection. What is here described is a fine example of what the
term means and how pupil choice really works in a practical
school situation.

      As is the case in any self-selection program of any
merit, there are here boundaries set within which such self-
selection takes place. Just as in individualized reading procedures involving self-selection an incredible number of books
must be constantly available, so in individualized mathematics
there must be a variety of materials of all kinds available at
all times. Otherwise self-selection becomes meaningless. I
have excerpted only a very brief account of these materials
and of how they were produced, but it can readily be seen
from even this brief discussion that providing adequate materials for such a program which allows free choice requires
much advance planning.

      This excerpt seems to me to speak well for itself, so
I shall make only one brief comment. You will note as you
read this passage that in two of the classrooms in which
IMCP was introduced, the teachers felt more secure if they
used the program with only half of their class, and both
teachers chose to introduce the self-selective procedures to
that half of their class which was achieving better. I understand the actions of the teachers completely and sympathize
with the feelings that prompted them to make these decisions
but I feel compelled to say that I disagree with their procedure. In the first place it is really easier in the long run
to apply such procedures to an entire class than it is to have
two different approaches operating simultaneously in one
classroom. Secondly, I am always inclined to believe that
if we are going to make changes in instructional procedures
we should make them first in the case of pupils who are not
succeeding as well rather than with those who are apparently
flourishing under the old procedures.

The second thesis excerpted is by Dr. Allen K. Neufeld and is closely related to the Bartel thesis we have just discussed.   An improved version of IMCP was used in all grades from the third to the sixth.   Professor Neufeld wished to ascertain what effect such a self-selection program had on children of different types of personality patterns and different intelligence ratings.   Somewhat in support of the proposition I advanced above in advocating the use of all the children in a classroom where experimental innovation is to be used, rather than a selected few, Neufeld found no statistically significant differences in the way in which the varying types of personality and intelligence affected success in the self-selection program.

Dr. Neufeld is now an associate professor on the Faculty of Education at the University of Alberta, where he takes an active part in various roles in connection with the Annual Invitational Conference on Elementary Education, of which the 1970 version deals with "Individualized Curriculum and Instruction. "

ACHIEVING INDIVIDUALIZED INSTRUCTION

Abraham Kaplan

It may seem to be a contradiction in terms to speak of achieving "individualization" in a "school. "   Yet, if we are willing to forego some of the accepted, time-honored practices, we can create an environment geared truly to meeting the needs and interests of individual pupils.

In New York City, through a variety of approaches, we have been attempting to meet the special needs of inner-city children.   We have tried smaller class size, added services, more attractive and more familiar materials, increased auditory and visual impact through multimedia, and broadening of cultural horizons.

While all of these approaches have been helpful to a degree, a significant cause of student failure often seems to be the group-oriented climate in which the conventional teaching takes place.   Traditionally, the organization of schools has been geared toward the group and not the individual.

Children may leave home as individuals--yet, from the moment they enter a school bus or a school door, they are organized and treated as parts of a group.

Teaching methodology as developed by Herbart in the
19th century, and which still guides much lesson development
today, was the response of a school required to deal with
faceless crowds.   This method requires that a lesson include
the following elements: motivation, aim, apperceptive recall,
logical development, summaries both medial and terminal,
and generalization and application.

Teachers are familiar with the following ramifications
of this oft-repeated and somewhat exhausted classroom process.   First, the teacher induces the class to listen as she
elicits the aim.   If she has prepared well, she ties the aim
to the experience and the background of the children.   She
devises a clever motivation to convince the pupils that they
ought to pay attention to her.

Then she asks a series of questions in order to achieve
her aim.   In accordance with the Herbartian prescription,
each child answers one or more questions.   Sometimes several children must be called on before the answer the teacher
seeks is forthcoming.   Eventually, students are expected to
weave all the questions and answers together in order to arrive at an understanding of the concept the teacher had set
out to teach.

By the time the fifth question has been posed, the
teacher usually has lost half the children, who now are
plunging to the bottom of the achievement curve.   They fidget
and squirm.   They whisper.   They play with things on their
desks.   Their eyes wander or they try desperately to get
them back to the business at hand.

This happens because the children couldn't read the
teacher's mind.   They didn't know in advance where the
teacher was going.   They may have missed the meaning of
one crucial word just as a pivotal question was being raised.
This is the general pattern of what happens when teachers
themselves take over--not, of course, with the intention of
dominating but with the assumption that they must drive children to a controlled goal or concept at all costs.

If we are committed to individualization of instruction,
however, it is evident that we must be prepared to seek new

approaches that do not rely so heavily on outworn strategies which invisibly pit the teacher against an inert mass of children. As Jerome Bruner has pointed out, the child learns most effectively when he discovers the concepts basic to the total curriculum through his own explorations and experimentation and through the manipulation of the tools of learning.

How may the school day be structured to achieve this? Under a program of "classroom decentralization" and individualization, it is proposed that the classroom be organized into simultaneous learning laboratories, thus breaking the pattern of a teacher who acts upon an audience of students.

The teacher will not develop a lesson according to the Herbartian "lesson whole." Rather she will participate with the students in work-study discovery centers in language arts, mathematics, social studies and science. Pupils will learn by manipulating, probing, researching and exploring as they move about these centers and discover projects. It is not to be expected that children will discover the algorism for division or the concept of the location of a point in space (latitude and longitude). These should be taught, but the teaching act need not be labored or total.

Models may be set up for children to redo what the teacher has taught (by manipulation and experimentation). Children should feel free to question one another and the teacher as they redo the demonstration and solve similar problems posed by the teacher, their peers, or by themselves. With this general pattern in mind we are prepared to understand the flow and movement of daily activities.

When the school opens in the morning, children will go directly to their classrooms, which will be open and ready for them. Much preplanning will provide children with a sense of purpose that will lead them to enter their rooms, put away their clothing and promptly turn to their appointed tasks.

Some pupils will attend to classroom chores such as watering plants, setting up the library corner, preparing materials, and gathering science equipment. Some will help their assigned buddies with reading, mathematics or social studies. Some will prepare dialogues for use in historical or literary dramatizations. Others will make the transparencies and overlays needed for social studies reports. Some

will peruse texts, almanacs and encyclopedias to discover
solutions to problems.   Some will prepare the mathematics
charts or games that their classmates will be using.

The important point here is that each child will find
the classroom to be made up of multiple points of interest,
a kind of organized "grab-bag" which invites interest and offers variety.   Some children will be at the mathematics work-
study center, using the pocket charts, splints and the abacus
to discover the solution to problems involving subtraction
with exchange.   Some children will experiment with rulers,
tapes, clocks and scales as well as with quarts and pints to
discover the mathematics of measurement.

In the library corner, there will be many books as
well as a flannel board, puppet stage, drawing paper, miniature furniture, trees, animals and figures for pupil use in
dramatizing or illustrating favorite scenes.   Less precocious
children will feel relaxed enough to browse, draw, prepare
flannel board or three-dimensional dioramas.   They will not
hesitate to ask a buddy for help with a word.   They will
keep envelopes that include the words they must learn.

In the writing corner will be a variety of available
themes for children to develop--a story for the class newspaper, a trip taken, a haiku about spring, a letter of thanks,
a get-well message.   The child may choose whatever theme
he wishes to write about.   In the art corner are materials
to develop paintings for the bulletin board or scenery for the
class play.   Paper, wool and ceramic materials are at hand
for the children to use creatively.

The teacher and class will decide on a great variety
of offerings and activities as well as on the rules of courtesy
and group behavior they will observe.   Perhaps quiet, relaxed socialization will take place.   Perhaps children will
elect to play a game of checkers or of dominoes.   Not all
activity need be academic, intense and controlled.   Various
types of play and seemingly random kinds of effort often are
the formats in which consolidation of learning is taking place.

The teacher will confer with individual pupils to clear
up reading difficulties, to assign reading worksheets, to rein-
force learning and to maintain personal contact.   She will be
finding answers to questions such as: Why does Timothy
seem to be troubled?   What's preventing the children in the
mathematics corner from solving problems posed?

Pupils can stop to review necessary facts once learned but now forgotten.   For example, in the mathematics work center there should be charts of number facts in the form of incomplete mathematics sentences.   For example: $8 + ? = 48$; $8 + ? = 15$.   Clipped to the bottom of each chart may be an answer key with the invitation: "Turn me over to check your answer. "

As a balance to this work-study pattern, the better part of the afternoon from day to day should be spent in explaining to the class the new problem posed in each of the work-study centers--in mathematics, social studies, language arts and science.   At this time there should be ample opportunity for ironing out problems, setting mutual goals, attracting further interest, and eliminating unworkable practices.

The following list of suggested materials and approaches in a model mathematics work-study center will indicate some of the possible resources that are available to children:

(1) Materials.   Hundreds frame, discs, pocket charts, abacus, fractional parts, squared materials, etc. ; liquid-measures such as a gallon, a quart, a pint, etc. ; and other containers of different shapes and materials.

(2) Statement of Problems.   Dittoed sheets for practice and self-evaluation are provided.   Suggested problems (in question form) may be kept on an attractive bulletin board until the teacher and class are sure they have mastered concepts.   Guide questions should be at hand to help the children experiment to discover solutions.   Children should be taught to use manipulative materials and to think through problems.   They should be trained not merely to accept an answer from a classmate, but to be able to "explain the answer. "

(3) Mathematics We Have Learned.   As skills are mastered, they should be listed on a chart.   Each child keeps a similar record in his notebook.   A file box should be at hand to keep dittoed materials as reference and as a source of supply for those who need it.

(4) Number Games for Practice.   Mathematics processes, lotto and dominoes, thinking cards, and flash cards; and arithmetic in daily life materials: price lists, newspaper ads, etc.   These materials should be neatly stored in

files and carefully and thoroughly taught before children can
be expected to use them independently.

Needless to say, a description of a work-study center
is not at the same time an account of a carefully programmed
school day. The latter can result only after the particular
resources of a school are directed to such a program. This
article has attempted only to establish general theory and
guidelines, which are by no means original or visionary.

Programs carrying out these principles are in effect,
in varying ways, in a number of schools. A broad program
is underway at the School University Teacher Education Center (SUTEC) located at P. S. 76 Queens. Visitors are invited to see the program in operation. In launching the program at P. S. 76 a group of ten teachers initially introduced
the pattern. At present, work-study procedures in whole or
in part are used in all classrooms. Supported by extensive
resources provided by the school, Queens College and the
community, the program has demonstrated a great potential
for developing forward-looking and significant patterns of individualized educational practice.

BREAKING BARRIERS BY INDIVIDUALIZING

Marilyn Jasik

Let us look at some classrooms in an ethnically integrated urban public school. The problems teachers face and
solve here have a degree of commonality with problems all
teachers face. Public School 76 has a certain uniqueness.
It houses SUTEC, a cooperative program in teacher education.

"If nothing else good happens this year, it won't matter, because I made it with Rosalie." These words are from
second-grade teacher Eleanor Beers. "Rosalie counts, measures and computes, but before she couldn't even name the
number symbols. She didn't know a 1 from a 20!" Admit-
ting that she was ready to give up many times with this quiet
girl who had been a "non-reader, non-speaker, non-every-
thing," Mrs. Beers persisted, nonetheless. Standard methods
had failed. An assortment of manipulative materials--beads,

discs, blocks and number puzzles--was ineffective.   Written
attempts were equally unsuccessful.   The idea of number, to
say nothing of the written symbol associated with that number, seemed beyond Rosalie's comprehension.   Looking for
clues proved to be a problem.

"I finally picked up what you might call negative or
reverse clues.   As I watched, I decided what would not
work, " Mrs. Beers explained.   "I ruled out large group or
small group activity, working with another child, manipulative
materials, pencils, papers, dittos, distribution of milk or
cookies, books, storytelling, use of the tape recorder or
overhead projector.   In fact, I ruled out everything I had
ever done with another child in the past. "

One day, in what she describes as a "last-ditch" effort, Mrs. Beers turned the class over to the student teacher, took Rosalie by the hand and went into the halls.   'One,
two, three, four, we marched and I counted ... something
clicked.   We counted steps; we counted doors; we counted
bulletin boards; we counted asphalt floor tiles.   We counted
everything!   One of the teachers passing by gave me a
quizzical look.   However, I sensed that I was on the right
track, and by December the light started to come through. "
Watching a new, lively Rosalie in her math groups, solving
problems on a rexograph sheet, one could hardly imagine
that she was once described as a silent, nonparticipating
child with sad, tearful eyes.

"Larry was a boy who could not learn his number
symbols ... he was verbal, loved stories, could count but
the idea of 'four' simply did not go with the symbol '4, ' no
matter how hard we both tried. " Sybil Wishner, first-grade
teacher, explained that she had noted the boy's ability in
handling puzzles and had a hunch that this tactile sense might
be utilized to further his perceptual learning.   But how? The
idea of making gravel numbers was an outgrowth of a remark
about commercial three-dimensional numbers made by a fellow teacher in a graduate course.   This triggered her imagination into inventing the gravel numbers for Larry.   Mrs.
Wishner believes that it was actually being involved in making the symbols himself which affected the boy's learning.
She then adapted the procedure for Bobby.

We watch six-year old Bobby play with and maneuver
fine gravel and liquid glue to prepare large three-dimensional
letters.   When they dry, he runs his fingers over the letters

and says them to himself.   Later, he arranges them in order
to make a word and copies it into his notebook.   "Bobby is
really learning now, " says Mrs. Wishner, "but until I tried
the number experiment with letters, I was not getting through
to him.   Now I have another problem--everyone wants to
use the gravel!"

INDIVIDUALIZING ARITHMETIC INSTRUCTION

Helen Redbird

        The United States' system of public education is based
on individual likenesses, not on individual differences.   The
idea is that we are alike enough that we can be educated together.   Individual likenesses is still a good concept to use
in the educational process.   It should not be considered as
opposed to individual differences, but as another concept in
arranging for learning in the classroom.   The real question
to me, as a teacher, is which one is the better concept for
a given learning situation.

        One of my basic assumptions in arranging for the
children's learning is that so-called average or normal children are capable of learning anything that they wish to learn
or that we might wish for them to learn.

        In addition to this basic assumption, there are four
questions that I will consider in describing the two-year
mathematics project completed in June 1963 in the Campus
Elementary School at Oregon College of Education, Monmouth,
Oregon:

        (1)   What specifically do you want the individual or in-
dividuals to learn?
        There were 32 fourth-grade children in the particular
group, and one of the specific areas they were to study was
arithmetic.   They were to learn about the usual things taught
at this stage of arithmetic instruction.   Topics considered
were numbers, addition, subtraction, multiplication, division,
money.   Roman numerals, measurement, familiar fractions,
estimating, and problem solving.

        (2)   Can the individual learn, and to what extent can

he learn?
        After considering all of the records and doing further
testing, it seemed that the group was average and above-
average in mental ability, as determined by standardized
measures.   The IQ range was from the low hundreds to above
150.   This would indicate that the children could learn at a
good rate.

        (3)   If the individual can't accomplish the learning,
why can't he accomplish the learning?
        One may well begin to wonder why there was concern
for this group.   It would seem, on the basis of a surface
descriptions, that the majority of the children would be able
to master the 11 areas of arithmetic instruction for this
grade level.   This was not true of the majority of the group.
This meant a search for reasons and explanations other than
poor preparation by previous teachers.   Some of the explanations were as follows:

        A.   Achievement was low in all areas.   Over 50 per
cent of the group had serious reading difficulties and 25 per
cent of the group was so retarded in reading skill that they
took all of the time of the special education teacher that was
assigned to the Campus Elementary School.   The reason that
the children qualified for all of the special time was their
high mental ability and low reading achievement level.
        B.   Three of the children had speech difficulties and
did not like to participate in oral communication.
        C.   One of the children could not write either manu-
script or cursive.   He had no discernible physical defect to
account for his writing inability.
        D.   One-third of the class was young, still eight
years old.   One of the children was taken out of the Gifted
Child Seminar because of his immature behavior.   He spent
more time crawling on the tables and under the furniture
than was generally expected typical behavior for this one-
third of the class.
        E.   Official psychiatric diagnosis indicated that one of
the children was brain-damaged.   He had great difficulty remembering things and keeping things in sequences.
        F.   There was really only one serious behavior prob-
lem, and this one was quite serious.

        (4)   What am I prepared to do, as a teacher, to help
him accomplish the learning?
        All of the three previous questions have to do with the
learner, but now the teacher comes into the picture.   The

following items were considered by the teacher and implemented by the teacher:

    A. Method: The method decided upon was to individ-
ualize the arithmetic instructional program to its ultimate
conclusion. Every child at a different place at a different
time.
    B. Materials: The materials used in the project
were the Seeing Through Arithmetic series by Scott Fores-
man; the Mathematics Enrichment Program Learning series
by Harcourt, Brace, and World; and any instructional device
that worked.
    C. Instructional Staff: The instructional staff in-
volved with this group was great in number and in deed.
There were student teachers, seminar students, special staff
from special education speech and reading clinic, psychia-
trists, psychologists, and the regular teacher in the self-
contained classroom.

Conclusion

    The advantages gained from this approach were two-
fold. The children learned arithmetic and they learned to
read. No child at the end of the two-year period was reading below a 6.3 test level, and no child was below a fifth
level in arithmetic testing. The greatest limitation in the
project was the number of papers and concepts that needed
constant checking. The time required was an unreasonable
amount of about two hours each day. In summary, the basis
for individualizing the arithmetic instruction or any instruction is expediency and efficiency in learning for both the
learner and the teacher. With this particular group, it
seemed hopeless to try any grouping as a means of individualizing the instruction in the last part of the second year.

INDIVIDUALIZED ARITHMETIC--AN IDEA TO IMPROVE
THE TRADITIONAL ARITHMETIC PROGRAM

Walter L. Whitaker

    The following program of individual teaching has been
tried and proved successful in the sixth grade in elementary
school in Culver City, California.

Materials

One of the keys to success in attempting this approach
is a wide range of source material.   Such material will give
each child an opportunity to actually learn at his own level
of achievement.   These materials should have a three- to
five-year range of difficulty.   This range will vary depending
upon class and grade.   Textbooks are many and varied with
different format and authors.   There are advantages in having
more than one series to use in the program.   Two to five
copies of each text for each grade level will provide sufficient working copies.

Workbooks, both old and new, in single copies provide
a vast source of reinforcement work when needed.   In both
textbooks and workbooks it is necessary to have answer
books.   In addition to the basic source material, there are
many books of the storybook type about numbers which can
be used as browsing readers.   These books offer enrichment
in the form of stories, games, and activities.

Presenting the Plan to the Class

Individualized arithmetic may be carried out with an
entire class or may include only a segment of that class.
To start, the best students should be put into the individualized work first.   This group may be given a long assignment
to begin with as a fundamental review.   As each child completes this assignment, he is placed in material equal to his
achievement and proceeds individually.

Once several children are launched solidly on a program of individual work, the balance of the class may be
gradually incorporated into the program.   It may take three
weeks for the entire class to be working in this program.
The slowest children will be the most difficult to start because they will not have the security of the group.   These
children need reassurance.   The arithmetic assignments
should be made easy enough so they can achieve some success.

If the entire class is put into individual work and it is
a sixth grade, the range of learning may well be 3.6 to 7.6.
Initially this program will create some confusion, but as the
child realizes he is learning at his own rate of speed the
group relaxes and a good learning situation ensues.

Teaching the Skills

Each child has his text or workbook, an assignment card, and a piece of graph paper. When he has completed the assignment he goes to the answer book and checks his answers. The score is plotted on the graph paper. Each error is analyzed by the child and, if he can, he writes why he missed the problem. When all this is completed he places his name on the appointment list which is on the blackboard. While waiting, he goes on with spelling, social studies, etc.

The teacher, during the arithmetic period, takes the names from the appointment list for the child's individual conference. As the teacher checks the work, a new skill may be started, a reinforcement (drill) assignment made, or some form of enrichment may be given.

If the child finds he cannot proceed with his assignment, he raises his hand, and this indicates that he needs help immediately. The child who raises his hand is taken before the next child on the appointment list. The teacher must be familiar with the skills both above and below grade level. From time to time some of the children will be ready for the same skills. At such time, this group may be introduced to the new skill and each child again proceeds at his own rate.

Checking Achievement

There are several ways to check mastery and achievement:

Achievement tests in the texts used.
Each text has progress tests which provide a superficial check of the child's progress.

Oral questioning.
Each time a conference is held an oral test occurs, for in this situation the teacher has the prime opportunity for determining true understanding. By careful questioning it can soon be determined how much a child really understands. Those oral conferences offer the best opportunity to ask "why" about the concepts the child is using.

Group testing:

Occasionally a group test is presented to the entire class for analysis to see the overall progress.

Any or all of the above ideas may be tried to upgrade the arithmetic program.   Careful planning is imperative. The following steps suggest some ideas: (1)   analyze the range of achievement from a standardized test; (2)   prepare specific plans relating the text material to each child on the basis of data in his cumulative record; (3)   arrange a meeting with parents to inform them of the program and plans; and (4)   inform the children several days in advance of the first day when the program is to be instigated.

An individualized program is different to some degree. This difference requires careful orientation of parents and children with the cooperation of the school administration. Not every child is seen every day.   Some may work three or four days while others will visit the teacher each day. To operate this program successfully, the teacher must deal with the children as individuals, not with a group at a particular grade level.   Each teacher who tries these ideas will need to vary the basic approach to meet the needs of the particular class.

CONCLUSION

The procedures presented here are basically a matter of organization with an increase in the content of the arithmetic program.   The intent is to show much of the detail involved in this particular approach to arithmetic in the classroom.   It is felt this new approach can offer each child a better education in terms of his abilities and interests without requiring a major change in the organization of the total school.   It can also be done with the materials usually found in the average elementary district.

INDIVIDUALIZED INSTRUCTION
IN ELEMENTARY SCHOOL MATHEMATICS

Elaine V. Bartel

From a comparison of individualized programs in reading with current individualized programs in arithmetic, and from an awareness of the frequent apathy of pupils in arithmetic classes in the intermediate grades, has grown the desire to study the feasibility of combining a program in new mathematics with a program of individualized instruction in the elementary school. The term "individualized instruction" as used in this study differentiates between existing programs called individualized instruction in arithmetic but in which programs no self-selection actually takes place, and programs in which each child actually seeks among the materials made available for him, makes his selection, and then goes to work at his task on his own rate, by himself or in a group of his own choice. It is in this latter sense that the term was used for the purpose of this study.

Development of the Project

The Individualized Mathematics Curriculum Project was begun at the University of Wisconsin in the spring of 1964. The purpose for which this project was undertaken was the preparation and implementation of an individualized program in elementary school mathematics in which pupils could select not only the topics they chose to study but also the materials they wished to use in working with these topics. The inclusion of the self-selection element in such an individualized program necessitated the selection and cataloging of a library of textbooks and enrichment materials to be placed in classrooms using the project. After an extensive review of texts and materials, and keeping in mind the necessity of providing a rich variety of materials while keeping the expense of such a library within reasonable budget limits, it was decided to include, in a fourth-grade classroom mathematics library, a quantity of textbooks and other enrichment material.

During the summer sessions of 1964, a team of eight graduate students who were involved in new mathematics programs at the University of Wisconsin worked together to plan the topics, or strands, to be included in the project and to prepare the eight strand booklets which the experimental groups were to use.

This group felt that all major concepts of fourth-grade mathematics could be included under eight basic strands. Since each pupil was to be free to select any strand or topic, it was necessary that each strand include its own prerequisite concepts, so that no one strand would have to precede any other. Within a strand, however, it was the general rule to proceed from simpler concepts to more complex topics, where such distinctions could be made.

The eight strands included in the project were: sets, geometry, numeration, rational numbers, problem solving, measurement, addition-subtraction, and multiplication-division. Each pupil involved in the experimental study was given a copy of each strand booklet. To facilitate for pupils the task of locating references pertaining to a chosen topic, the arithmetic books in the classroom library were coded according to publisher and grade level, and this coding was placed on the binding of each book with heavy felt-pen markings.

Wherever applicable, each topic within a strand began with references from third-grade texts and proceeded through sixth-grade texts. Since not all topics are covered at all these grade levels, it was not always possible to follow such a procedure, and it must be kept in mind that the strand itself was constructed in ascending difficulty levels of basic concepts.

The initial writing, revisions, and final preparation of these strand booklets were completed during the eight weeks of the university summer sessions, so that the only task remaining in the preparation of materials was the design and assembly of the pupil folder. First, a chart of numbered spaces was stapled in the left cover. When a pupil felt that he had completed a topic within a strand to his own satisfaction, he colored the space which contained the topic number, using a color code which enabled one to see the approximate date and sequencing of various topics studied. Stapled in the right side of the folder were eight sheets which corresponded to the eight strands of the project. On each sheet the pupil could find a complete list and order of topics included in the strand, and this was the listing he consulted as an aid in selecting a strand as well as a level, or topic, within a strand in which to work.

A supply of the daily worksheets was kept in the experimental classroom, on the portable cart containing the library of reference books.   Pupils did all computations, corrections, and scoring on these sheets which were kept in the folder and used as a basis for the teacher-pupil conferences.

A pilot project was initiated in a fourth-grade classroom at Watertown, Wisconsin.   This class used all the materials of the program from October 1964 to June 1965. Pre- and post-tests were administered to this group, interviews were conducted, and attitude questionnaires were completed and analyzed.   It was felt that such a classroom could serve the purpose of providing feedback information necessary for the continuous revision and subsequent improvement of the program before it was implemented in the experimental and control classrooms for the major research portion of the study.

During the first semester of the 1964-1965 school year the writer kept in close contact with the teacher at Watertown.   On the basis of these contacts, and from insight gained from frequent visitations of this classroom, many changes were made both in the mechanics of implementation as well as in the materials of the project.

After the Watertown class had been involved in the individualized program for several months, it became increasingly apparent that the total lack of testing material represented a major limitation of the project.   The classroom teacher felt she needed a more definite basis for encouraging each child than his daily worksheets could afford, and it was felt that the pupil himself could benefit from a test of his understanding of basic strand concepts.   Such strand tests could serve as diagnostic as well as mastery instruments, and it was with this twofold purpose in mind that the eight strand tests were constructed.

Each of the eight tests consisted of four-option multiple-choice items, and a geometry construction test was prepared in addition to this basic set.   Since the tests were to be administered during a regular arithmetic class period of 45 to 60 minutes, it was felt that the multiple-choice design would most easily lend itself to a wide sampling of concept understanding in a minimal time span.   The preparation of these strand tests was completed in February of 1965, and the entire set of tests was ready for use when the experimental program was begun in the Brodhead School.

Brodhead is a small city located 35 miles south of Madison, with a population of 2400. The Brodhead Elementary School has an enrollment of 830 pupils, and includes kindergarten through eighth grade. Agricultural and factory work make up a large percentage of the occupational categories of the parents of these children.

In the early stages of the inservice mathematics education program at Brodhead, the materials of the Individualized Mathematics Curriculum Project were used quite extensively as a basis for introducing to these teachers basic concepts and materials of the new mathematics programs. In addition to biweekly lectures by university personnel on key topics in new mathematics, these teachers used the strand booklets, pupil folders, and library of textbooks to study topics of particular interest to them, at the grade level of their own choice. Use of these materials served to thoroughly acquaint the teachers with the mechanics of the individualized program as well as its scope and potential for experimentation.

In January 1965, two fourth-grade teachers were selected to use the individualized program in their classrooms for the remainder of the school year. In one of these rooms the entire class of 26 pupils was to be included in the individualized program. This fourth-grade class is referred to as the experimental class for the pupil achievement portion of this study.

The teacher of the other fourth-grade class used the individualized program with approximately half of the pupils in her room. These were the pupils who in this teacher's estimation were best suited for a program of independent study. In general they were the top achieving pupils in arithmetic. This group was composed of 14 pupils. This class has been named the split class to distinguish it from the other experimental class.

These two classes were used in obtaining data concerning the materials of the program and their use. They were also included in this study to determine teacher and pupil attitudes toward the many phases of the program. The experimental class and the split class are referred to as the experimental group when both are included.

The final question of the study upon which conclusions concerning feasibility were based was: "Are pupils able to make reasonable progress while participating in such a program?" For this pupil achievement phase of the study, the pupils in the experimental class were compared with the pupils in a control class.

The question, "Do the teachers generally find the program teachable, and do they like it?" was also posed to determine feasibility of the individualized mathematics program. Close contact was maintained between the writer and the participating teachers throughout the study, and each teacher was extensively interviewed during the final week of the study. The outstanding impression resulting from these interview sessions was the similarity of comments made by these teachers. There was agreement, for example, on the following points:

(1) The individualized mathematics program permits each individual to advance at his own rate.
(2) Pupils develop independence and self-confidence while participating in the program.
(3) The program helps children develop skills in using reference materials.
(4) Children develop a sense of responsibility in checking their own work and recording their progress.
(5) The pupils are happy and relaxed.
(6) The program provides the opportunity for pupils to move about the classroom freely to obtain materials, seek help, etc. releases tension.

One teacher reported that it was easier for her to teach arithmetic by using the individualized program. Both teachers expressed the desire to use the program throughout the next school year with their entire classes. However, they raised some questions concerning certain features of the program. For example, they expressed concern over their task of assigning a letter grade to pupils working in the experimental program. These teachers also agreed in their comments suggesting that the slower pupils were the cause of concern in several instances. For example, it was mentioned that the poor reader is at a decided disadvantage in a program where comprehension depends upon ability to read the material independently. It was also mentioned that the slower pupils did not keep accurate records. Both teachers

commented that the pupils who in their estimation needed
help were the ones who were most reluctant to seek help.
The teachers also felt that they did not have adequate amounts
of time to spend in teacher-pupil conferences.   Both teachers
felt that the program was teachable, however, and generally
found ways to overcome the problems unique to an instructional program of this type.

Several data-gathering procedures were used to explore pupil attitudes toward the individualized mathematics
program.   Responses to the items on the attitude inventory
serve to illustrate the general reaction of pupils in the experimental group.   A questionnaire was used as the basis
for an interview with each pupil in the Brodhead experimental
group.   These interviews were conducted by a university student.   It was felt that an evaluative instrument of this nature
would permit pupils to interpret and respond to the items in
such a way as to lend insight into their true feelings about
various aspects of the program.

The responses on both attitude inventories confirmed
the numerous casual remarks made by the pupils throughout
the course of the study, indicating extremely positive attitudes toward this method of studying arithmetic.

Summary

Within the basic framework of this study certain questions were posed for the purpose of lending insight to the
various facets of the feasibility of a program of this type.
Subjective as well as objective analyses were applied to the
data and observations of the project.   On the basis of these
observations and analyses it was found that the materials of
the study were used in a logical and meaningful way by
participating pupils; the children were able to adequately select materials as well as topics for study, although they became far more adept at this procedure after the initial
stages of the program; the participating teachers as well as
the pupils were generally quite positive in their statements
of attitudes toward the program.   Specific features of the
program which the teachers or the pupils did not like were
enumerated, and all test data analyses were reported for
determining whether or not pupils involved in the program
made reasonable progress.

In the analysis of covariance it was found that when

the Iowa Achievement Test was used as the dependent variable to measure arithmetic achievement there was no difference in achievement between pupils in the individualized arithmetic program and pupils in a traditional instructional program.   When the Concept Test was used as the dependent variable, however, pupils in the experimental program scored consistently higher than pupils in the traditional program.   This difference between groups was found to be significant beyond the .01 significance level.

PERSONALITY AND ACHIEVEMENT UNDER
INDIVIDUALIZED INSTRUCTION

K. Allen Neufeld

        There seems to be evidence that present curricular content and method may not be meeting the needs of many children.   It is imperative that appropriate curriculum projects be designed and implemented to provide bases on which curriculum decisions can be more adequately determined. The specific focus of this research is on selected characteristics of pupils that are related to achievement in a program of individualized instruction in mathematics.

        Extensive changes in mathematical content for the elementary schools have occurred during the last decade.   Some changes are closely related to acceleration, in that certain content is taught sooner than in previous times.   Some content which previously would have been classified as enrichment for some learners is now prescribed as suitable material for all learners.   The intersection of content among the various texts for a certain grade level is much smaller than previously.   An unlimited field for research has thus been opened in the area of content.   There are many indications that the content of mathematics, with its recent overall revisions may provide subject matter that is attractive to the problem learners focused upon by this study.

        Computation and problem solving, previously a large part of the whole program, now represent only a fraction of the content encountered by the learner.   There is great opportunity for children to find an area in mathematics in which they can succeed.   Even in computation, rather than being

confined to one pattern, a variety of algorithms are presented.   Contemporary texts present problem solving in a
variety of ways.   Some present several methods of attack,
while others present a solution method which emphasizes
reasons for a particular choice of operation.   These revised
points of view relative to two content areas provide rationale
for the belief that mathematical content now has features attractive to a larger number of students.   Besides these new
insights into traditional topics, contemporary programs present a wealth of diversified content.

Heyns [1end of article] states that "the school calls
on abilities that are highly correlated.   The range of skills
required in school is narrow and the child who achieves success in one skill is likely to be successful in another, similarly, the child who fails at one is likely to fail at another. "
The content of mathematics now features a wider range of
skills than any mathematical content previously taught.   The
chances that every child can find some area of special proficiency are increased.   The content factor can be an integral part in helping a child to gain new insight into his academic potential.

Methods to Meet Individual Differences

The values and learner-needs referred to at the outset of this paper almost directly imply that the teaching-
learning act must be concerned with individuals rather than
groups.   The topic of individualization is appearing increasingly often in professional literature.   Organizations such as
the National Society for the Study of Education and the Association for Supervision and Curriculum Development have
recently devoted entire yearbooks to this topic.

Weaver[2] summarizes numerous publications dealing
with differentiated instruction, many of which provide a
rationale for such an approach.   Some of these instructional
plans are too theoretical; they lack practical applicability.
In other instances, a revised procedure of instruction is
accompanied by a host of new problems, some as serious as
those answered by the new plan.

If society truly values the tenets of individual worth
and equal opportunity and recognizes the learner-needs of
creativity, responsibility, positive self-attitudes, and motivation, then there is need of a plan of individualized instruction

which will optimize learning, especially for those children whose learning is hampered by large group classroom procedure.

Statement of the Problem

The problem of this study is to determine the relationship of selected characteristics of pupils to achievement gain in elementary school mathematics classes using the Individualized Mathematics Curriculum Project (IMCP) approach.

Hypothesis Number One: There is no difference in personality characteristics among pupils of different levels of mathematical achievement gain.

Hypothesis Number Two: There is no difference in intelligence among pupils of the different levels of mathematical achievement gain.

Instructional Program

The Individualized Mathematics Curriculum Project (IMCP), the treatment used in this experimental study, is a major revision of the original IMCP of the Bartel[3] and Fredriksen[4] studies. This refinement by mathematics educators, mathematicians, and classroom teachers was undertaken during the two 8-week summer sessions of 1965 and 1966. In addition, the project was piloted in 20 classrooms at the third- through sixth-grade levels during the 1965-1966 school year. Many of the fifth- and sixth-grade students of this study were involved in this piloting for approximately two months. The classroom teachers involved in this study participated in an in-service education program related to IMCP just prior to this study. Since major organizational revisions were undertaken during the summer of 1966, the program presented to the experimental classes and their teachers contained many improved features.

A variety of materials is utilized by IMCP. The basic third- through eighth-grade textbook series of five different publishers account for a large portion of the printed content. In addition, three different series of enrichment materials are available for pupil use. The basic textbook series were chosen to represent a wide range of content and method.

The materials are organized under the following eight strands: numeration-number theory, addition-subtraction, multiplication-division, measurement, geometry, numbers-number systems, sets, and techniques for solving written problems.   Each strand is composed of from three to ten units, while each unit covers from three to 12 related mathematical topics.   The organization of the units and topics of IMCP is based on a selected mathematical sequence of the common content which is found in the basic textbooks and enrichment booklets.   While each strand category is treated in all the textual materials, individual unit and topical content is drawn from one or more of the variety of book and booklet series.   Each topic also is treated at one or more grade levels but only occasionally at all six levels.

To aid pupils in finding materials based on a particular content topic, cross-referenced index booklets are used.   All the references to the appropriate series, grades, and page numbers for a particular topic are contained on a single page of the coded index booklet.   The topics and units for each strand are also listed on separate charts to provide each pupil with a convenient scope and sequence reference.

The degree to which pupils are allowed to self-select materials is dependent on several factors.   As the program is being initiated and the organizational mechanics of IMCP are being learned by the pupils, selection is usually limited to three strands, namely, sets, addition-subtraction, and multiplication-division.   Early units in each of these strands outline mathematical content basic to most of the other strands.   Each pupil is instructed to choose a unit which seems to accommodate his knowledge level.   As the pupil reads textual material and works exercises for a particular topic in a chosen unit, he soon discovers his own capabilities in relation to the material.   One of three plans of action may be chosen.   If the material is much too difficult, the pupil may, in consultation with the teacher, select an easier unit. If the pupil feels that the material is approximately at his level, he may consult several references for each topic and study the material.   The presentation of a particular topic in one book may be more meaningful than that found in another book at the same grade level.   In this way he works through several references for each unit topic.   If a pupil chooses a unit which is much too easy for him, he is expected to work a few sample exercises throughout the unit. If he finds a minimum of difficulty, he can record completion of the unit.

The pupil's self-selection is somewhat limited in that a unit of content chosen must be completed with satisfactory results shown. An exception exists in that a unit which proves much too difficult may be exchanged for one of lesser difficulty. A chart to indicate completed units is kept by each pupil.

After the mechanics of the program are understood, and pupils have established some idea of their capabilities, the early restriction of three strands is removed. Retaining only the unit-completion restriction, pupils are free to select any unit of any strand.

The aspect of self-pacing has already been discussed to a certain extent. A slow pace may be chosen by pupils who feel they need to work many references of a particular topic to gain full understanding. Others can accomplish this achievement by working a minimum number of references. A pupil may also use spare time during the school day or the evening hours to work on the program, thus making decisions on his own chosen pace.

The teacher in IMCP spends a major part of his time in counselling with individuals with regard to selection and pacing problems which they may encounter. At times, the teacher may present to the complete class the features and organizational pattern of a potentially interesting unit or strand which few have undertaken. The pupils often appreciate the variety which a lecture to the whole class presents after several weeks of individual study. If the members of a small group are having common problems, a session of some remedial teaching may be planned. Most often, the meetings are between the teacher and a single pupil. During these times a teacher encourages and exchanges ideas with the pupil. The pupil may wish to discuss interesting findings, or to report on recently discovered relationships between strands.

Evaluation is an essential part of IMCP. Each pupil conducts a self-evaluation several times each day as he uses the answer key in the teacher's manual to correct his exercises. Each teacher has a file of unit tests which a pupil may call for in order to evaluate his mastery of the content in a particular unit. Strand tests are available for a more comprehensive evaluation of each of the eight major areas of content. The main feature of these tests is that they were constructed to evaluate the common content of a variety of text and booklet series.

Data Collection

In early November 1966, immediately prior to the
time that the classes were first organized on the basis of
IMCP, the Wisconsin Contemporary Test of Elementary
Mathematics (WCTEM), the Stanford Achievement (Arithmetic)
Test (STA), and the California Test of Personality (CTP)
were administered to all pupils of the fourth, fifth, and sixth
grades.   The Kuhlmann-Anderson Test of Academic Potential
was administered to the same pupils with the exception of
the sixth grade.   The scores of a previous testing were used
for the sixth grade.   Following a period of six months of
IMCP instruction, the two mathematics tests were re-admin-
istered to all pupils.

Results

The multivariate F-ratio tests indicated that there
were no significant differences between pupils exhibiting high
and low achievement gain on the composite of dependent variables of this study.

Only 21 out of a possible 324 cases were noted in
which the high and low groups exhibited significant differences
on certain personality characteristics.   Interpretation of these
overall results emphasized that there was a minimum of difference in total personality between pupils of the high and
low achievement gain groups.   The majority of differences
that did occur favored the high group at the fourth-grade level.
The few differences at the sixth-grade level favored the low
achievement gain group.   Prior to IMCP, 37 differences were
found, all at the fifth- and sixth-grade levels.   A directional
trend was evident in that from the fourth- to the sixth-grade
levels, significant differences in the pre-test data increased
while differences in the achievement gain data decreased.

With regard to the dependent variables of intelligence
and sex, the overall results indicated that there was no difference in intelligence and sex between the high and low
achievement gain groups.   The directional trend discussed in
the previous paragraph was also evident for these variables
with respect to the pre-test differences only.   Whereas 21
significant differences were found relative to the pre-score

analysis, no differences were found in the analysis of the
achievement gain data.

Bartel has suggested that IMCP was feasible to adopt
in the classroom.   Teachers and pupils, with few exceptions,
were satisfied with the program.   Any implications for the
regular classroom are based on the degree of individualization that is implemented in the regular classroom.   Whether
the plan of individualization exactly follows one of those previously reviewed or is patterned after IMCP is not important.
It is important, however, to realize that many features of
various individualized programs may be incorporated into the
day-to-day program of any classroom.   It is possible that
the results of this study may occur under other circumstances
of individualization.

The findings of this research, relative to the personality variables, would imply that there is little difference in
the personality structure of groups of pupils exhibiting high
and low achievement gain.   Since the fewest differences were
found at the sixth-grade level, the implications would apply
most strongly for this group of pupils.   There is a possibility, however, that the occurrence of these few differences
was related to the fact that at the sixth-grade level, a higher
percentage of the total class received individualized instruction than at any other grade level.

What implications for curriculum and instruction may
be found in the results pertaining to the personality variables?   Consider a pupil, in a regular classroom, who feels
a lack of personal worth.   There are pupils all around him
who are succeeding at curricular tasks which he is having
difficulty mastering.   It is possible that this pupil's inferiority might result in low achievement.   The same child receiving instruction individually may be less aware of inter-
class competition.   He may not be restricted to any one set
of curricular material.   His answers to questions posed by
his teacher may not be subject to the scrutiny of his peers.
If his feeling of inferiority is not put on display, there is a
possibility that his achievement might not be affected.

Early in the planning of this study, the value judgment
was made that the instructional methods of the classroom
should be compatible to the personality structure of the pupil.
There is need of research which will evaluate the compatabil-
ity of pupils to various methods of instruction.   The results
of the present study indicate relationships that exist between

selected personality characteristics and achievement gain in
the setting of IMCP. Would the results be similar in a procedure involving team teaching, ability grouping, or nongrading? An experimental study in which pupils were randomly
assigned to a classroom utilizing each of these methods or
the regular class-as-a-whole method might generate results
with relatively direct implications for the classroom with regard to personality-achievement relationships.

The findings of this research, relative to the intelligence variable, would imply that the mean intelligence of
pupils exhibiting high and low achievement gain is no different. What implications for curriculum and instruction may
be found in the results pertaining to the intelligence variable?
Intelligence scores have long been one of the factors in grouping class members for instruction. Teachers have generally
felt that, because a low intelligence group requires a different curriculum and method of instruction than a high intelligence group, the low intelligence group could not achieve as
well as a high intelligence group. An individualized program
such as IMCP allows pupils to self-select curriculum materials to satisfy their own needs. The results of this research may imply that, even though each pupil is working at
a level compatible with his own ability, low intelligence does
not limit a pupil's ability to exhibit significant gain in
achievement. There is reason to believe, however, that over
a longer period of time in individualized instruction, the
positive relationship between achievement gain and intelligence may increase. As the pupils of high intelligence utilize
more fully the acceleration and enrichment features of an individualized program, and as instruments are available to
measure more of the content areas which might be covered,
the relationship between achievement gain and intelligence
may well increase in comparison with the relationships reported in this study.

The problem of the present study was to seek relationships between the variables of intelligence and achievement gain within the context of a single treatment, IMCP.
There is a need for research which is designed to analyze
the interaction of teaching procedures relative to individual
differences.

An individualized procedure such as IMCP may allow
the pupils of high intelligence to master, in a short period
of time, the basic mathematical concepts as measured by
available achievement instruments. While the pupils of low

intelligence are taking a much longer time to master these
same concepts, the high intelligence group may be mastering
enriched content, or content well beyond their grade level.
It is possible that the lack of difference in intelligence between the high and low gainers is due to the fact that available achievement instruments are not designed to measure
the content which the high intelligence group have mastered.
There appears to be a need for the development of instruments to measure the widest possible range of achievement.

The finding that there were no significant differences
in the ratio of males to females between pupils exhibiting
high and low achievement gain would imply that sex is not a
factor in determining academic success in an individualized
program.   This result basically confirms the findings of previous research.   Subjective judgments by teachers have indicated that boys at the upper elementary grade levels tend to
be less academically inclined than girls.   If this judgment is
warranted, and if the boys feel academically inferior due to
the competition afforded by regular classroom procedures,
then perhaps an individualized program can alleviate some of
this inferiority felt by boys.

The results of this study are based on the relationship
of personality, intelligence, and sex to achievement gain in
an individualized program of instruction.   Implications of the
results are limited to the extent that the Individualized
Mathematics Curriculum Project was instrumental in promot-
ing pupil achievement.   Although some past research has
been done comparing IMCP with other patterns of instruction,
a need still exists for further research to make comparisons
among several instructional methods.   The present study
points up a need for the continuous evaluation of programs
of instruction, and the examination of characteristics of pupils related to achievement in these programs.

                             Notes

1.   Roger W. Heyns.   The Psychology of Personal Adjust-
      ment.   New York: Holt, 1958.

2.   J. Fred Weaver.   "Differential Instruction and School-
      Class Organization for Mathematical Learning Within
      the Elementary Grades."   Arithmetic Teacher 13:495-
      506, 1966.

3.   Elaine V. Bartel.   <u>A Study of the Feasibility of an In-</u>
      <u>dividualized Instruction Program in Elementary School</u>
      <u>Mathematics</u>.   Doctoral dissertation.   Madison: Uni-
      versity of Wisconsin, 1965.   [See the previous article
      for excerpts from this document.]

4.   Kathleen Stone Fredriksen.   "A Report of an Experi-
      mental Individualized Arithmetic Program."   Unpub-
      lished master's seminar report.   Madison: Univer-
      sity of Wisconsin, 1966.

V. INDIVIDUALIZING MATHEMATICS INSTRUCTION:
    2. FURTHER ILLUSTRATIONS

        In this chapter we continue our examination of descrip-
tions of ways in which mathematics instruction has been individualized. As we are made aware of the variety of different ways in which this has been done, it becomes more
practicable for us to assemble ideas that we feel are most
adaptable to our own situations.

James E. Bierden

        The first passage is taken from the 1968 University
of Michigan doctoral dissertation of Professor James E.
Bierden, who is now on the faculty of Rhode Island College
at Providence. It is easy to assume that he has found a
congenial niche there, as this institution has long been in the
forefront in developing and disseminating ideas about individualizing the instructional process.

        As was the case in the article about the Borel Middle
School in a previous chapter, the emphasis of the program
described here is on the use of behavioral objectives as the
principal way of evaluating the individual pupil's performance.
It will be recalled that a question was raised in my introduction to the chapter containing the description of the Borel
program of individualization about the desirability of having
the same level of behavioral objectives for all children; in
the Bierden plan behavioral objectives on each topic are differentiated into three levels.

        An approach worth thinking about is the whole class
introduction to each topic. This would, of course, be impossible unless behavioral objectives were varied as they are
here. However, one may question whether it is possible to
truly individualize instruction and at the same time keep the
whole class more or less working on the same topic. It

appears to me that it would be very difficult to do this without at least to some degree slowing down the most rapid and
the most diligent pupils or speeding up the process of completion of each topic or unit by the slowest learning pupils,
with the result that their grasp may not be as adequate as
would be desirable even with the more readily attainable objectives.   The question that I raise here might be answered
according to the philosophy of the particular teacher.   It is
a matter of balancing the advantages of keeping a class
somewhat together and allowing each child to pace himself
in accordance with his capacity and the degree of interest
that has been engendered.

I have tried to extract from this very fine thesis not
so much Dr. Bierden's explanation of his rationale, which is
extremely interesting, but rather some specific examples of
the materials used to carry on his program.   Since these
materials are largely self-explanatory, I shall not dwell on
them at length in this chapter introduction except to express
my admiration for the carefully developed nature of the
materials.

The findings and discussion of their meaning make it
clear that the program was successful in imparting the desired understanding and knowledge to the pupils and that the
method used was satisfactory and satisfying to all those taking part in the process.   The discussion emphasizes the fact
that there is always room for modification and improvement
in any instructional program that is vital and innovative.

Marian J. Patterson

The second selection in this chapter is by Mrs. Marian J. Patterson, a mathematics consultant for the Los Angeles City Schools.   It is a descriptive study of the use of the
computer to individualize instruction for fewer than 20 under-
achieving and disadvantaged pupils from two seventh grades.

Those who are impressed only by reports of carefully
controlled studies including a pre-test and a post-test which
are examined to determine whether there is a statistically
significant difference in either or both groups or between the
control and the experimental group, as far as either test is
concerned, will not find this excerpt to their liking, for it
includes none of these items.   It is just what it says it is,
a descriptive study.

In later chapters a considerable amount of space will be devoted to the role of the computer in the individualization of instruction. The mathematics program described here has been developed by Stanford University's Professor Patrick Suppes, head of that university's Institute for Mathematical Studies in the Social Sciences. His program has been given trials in many school systems, including major urban centers such as New York and Los Angeles. There is some difference of opinion concerning the success of this program in actually individualizing instruction and questioning the degree to which mathematics is learned more efficiently using individualized techniques. One difficulty is that the Suppes system constitutes a partnership between the classroom teacher and the computerized lessons. If the program does not yield the desired results it is easy to attribute this failure to either the classroom teacher or the computerized system, depending on the observer's individual interpretation. It has been my own personal observation that the in-service training and supervision of the classroom teachers involved has been inadequate. It would seem to me that better training and supervision would be absolutely essential to a fair test of the adequacy of any proposed plan of instruction.

The fact is that the computerized work is skillfully planned and organized so that there is a fine opportunity for learning in a good environment, which would most certainly include a classroom teacher who is adequately briefed and who, just as importantly, is in sympathy with the method used.

It should be noted that there is a considerable amount of provision for individual differences in the Suppes scheme of things but that there is practically no opportunity for self-selection or pupil's choice in any respect. To that extent, of course, this system does not fully comply with what has been described as the essentials of individualized instruction earlier on in this book.

Eugene R. Keffer

Professor Keffer is director of institutional research at Central Missouri State College at Warrensburgh. In 1961 he wrote an article for Arithmetic Teacher in which he described a plan for the individualization of arithmetic instruction that he had observed at the State College's laboratory school.

The plan, lucidly presented in the third portion of this chapter, is I think quite clear and therefore requires a minimum of comment or introduction. However, I should like to call attention to the fact that no specifically prepared materials, blanks, and so on, are required to carry on this program. This feature may make the plan more attractive to many classroom teachers, as well as to school administrators, than a more elaborate plan such as is described by Dr. Bierden or by Mrs. Patterson.

George C. Nix

The fourth item in this chapter consists of selected portions of a University of Alabama doctoral dissertation written in 1969 by Professor Nix of Georgia State University's Graduate Center in Columbus. Dr. Nix writes me:

"My interest in individualized instruction began some time ago. I kept reading and hearing about individualized instruction but not seeing it. Other teachers and I did not have either machines or programmed materials, usually because of a lack of money and fear by the teacher that she would not know how to use these things if she had them.

"I simply wanted to find out if a teacher could individualize her instruction without having these things, if she had only the normal textbooks and other basic materials."

Accordingly, Nix wrote a dissertation in which he reported a classically designed experiment. I leave it to the reader to decide whether the results justify the use of individualization, which was the experimental factor in his report. I do think that the factor of interest generated is of vital importance.

My comment about Keffer's description, in which I stressed the fact that it contemplated the use of only those materials that would already be available to the average teacher, certainly also applies to the description of this project. I think that it might well be argued that the future of individualized instruction in American schools may depend more on approaches such as the one described here than on computer-based, programmed, and other approaches which require a considerable investment of time and money in special equipment. Time alone will prove or disprove this judgment.

R. B. Thompson

While the last article in this chapter is quite sound, I have included it principally because it was published in 1941. This, I think, adequately supports the idea that individualized instruction is not just a passing fad, thought up in the last few years.

INDIVIDUAL DIFFERENCES IN SEVENTH-GRADE MATHEMATICS INSTRUCTION

James E. Bierden

The purpose of this study was to explore and develop a form of classroom management designed to aid in the provision for individual differences among students in a seventh-grade mathematics course. The two major variations from normal classroom procedures involved in the study were (1) the classroom use of detailed behavioral objectives related to the content of the course, and (2) a form of classroom management using a combination of whole class instruction and flexible intra-class grouping based on achievement of objectives.

Procedure

Two seventh-grade classes of the University School, the University of Michigan, were taught using the classroom management procedure during the 1967-68 school year. Detailed behavioral objectives were written at three levels, basic, intermediate and advanced, for each topic covered in the course. In its final form, the management procedure for each topic included (1) initial instruction of the whole class designed to meet the intermediate level objectives, (2) testing the students on these objectives, (3) assigning each student to a basic, intermediate, or advanced group depending on his achievement of the objectives on the test, (4) small group and individual instruction aimed at students' unachieved objectives, and finally (5) a second test on these unachieved objectives.

MATERIALS FOR A TOPIC*

TOPIC 8.   MULTIPLICATION AND DIVISION OF FRACTION-
AL NUMBERS

Textbook: Chapter 5, pages 215-244
General objectives:
1. To review basic concepts of fractional numbers.
2. To compare fractional numbers named by fractions and to be able to give equivalent fractions.
3. To multiply and divide fractional numbers named by fractions.

Reasons for studying this topic:
1. There are many situations in which fractional numbers are needed to represent ideas or to compute quantities.
2. Studying concepts of fractional numbers also helps us review the concepts of whole numbers.

Section 1:   Number Pairs and Fractional Numbers

8 - 1 - 1 - I/216
   To tell what is meant by a fractional number.
      C: Answer includes an idea the same as: "A fractional number is a number that can be named by a fraction with a whole number as the numerator and a counting number as the denominator."

8 - 1 - 2 - B/217, 219
   To recognize fractions equivalent to give fractions.
      E. For each set of fractions, put a loop around each member which is equivalent to the first member.
      1. (2/3, 14/21, 8/12, 12/15, 18/27, 20/28).
      2. (5/8, 20/32, 30/50, 15/24, 25/40, 40/64).
      3. (7/2, 15/3, 21/6, 42/10, 49/14, 56/18).
      C: 80% correct.

8 - 1 - 2 - I/217, 219
   To write fractions equivalent to given fractions.

---

*Editor's note: the passages that follow are designed to resemble teachers' actual working papers.   Internal references (i. e., to chapters, page numbers, "sections," etc.) are retained as purely illustrative, without relating to anything in the present work.

E: 1. Find the replacement for n which makes the second fraction equivalent to the first.
   a) $2/3 = n/18$ b) $0/5 = n/10$ c) $7/8 = 56/n$
   d) $4/4 = n/12$ e) $12/1 = n/1$ f) $20/28 = n/7$

 2. Write three fractions equivalent to each fraction below.
   a) $3/8$ b) $10/7$ c) $0/5$

C: 80% correct

TOPIC 8	ASSIGNMENT SHEET	NAME _____
OBJECTIVE	PHASE I	PHASE II
8-1-1-I	Obj.	Obj.
8-1-2-B	-------	supp.
8-1-2-I	p. 220, 6-16 even	p. 220, 7-17 odd
8-2-1-B	-------	p. 223, 5, 7, 9
8-2-1-I	p. 223, 4-14 even	p. 223, 11, 13, 15
8-2-2-I	p. 223, 16, 17	p. 223, 18-20
8-2-2-A	-------	p. 223, 21-23
8-3-1-B	-------	p. 226, 1, 3, 5, 7, 9
8-3-1-I	p. 226, 2-34 even	p. 226, 11, 13, 15, 17, 19
8-3-1-A	-------	supp.
8-4-1-B	p. 137, 29-32, 57a-c	p. 137, 33-36, 57c-e

[and so on]

TOPIC 8	TEST 1	Name _____

8-1-1-I   Tell what we mean by a fractional number.

8-1-2-I   Find the replacement for n which makes the second fraction equivalent to the first, and write them in the blanks.

 1. $\dfrac{5}{7} = \dfrac{n}{21}$____   2. $\dfrac{0}{4} = \dfrac{n}{8}$____   3. $\dfrac{5}{6} = \dfrac{40}{n}$____

4. $\dfrac{3}{3} = \dfrac{n}{n}$ 5. $\dfrac{n}{9} = \dfrac{4}{n}$

Write three fractions equivalent to each fraction below.

1. $\dfrac{3}{5}$ _____   2. $\dfrac{17}{8}$ _____

3. $\dfrac{8}{8}$ _____   4. $\dfrac{0}{25}$ _____

8-2-1-I   Solve the following sentences. The replacement set for n is the set of fractional numbers.

1. $\dfrac{4}{7}$ of $9 = n.$ _____   2. $\dfrac{3}{4}$ of $\dfrac{7}{15} = n.$ _____

3. $\dfrac{3}{8}$ of $80 = n.$ _____   4. $\dfrac{9}{5}$ of $\dfrac{12}{7} = n.$ _____

5. $\dfrac{1}{5}$ of $\dfrac{4}{25} = n.$ _____

8-2-2-1   Solve the following problems and write the answer in the blanks. Write your answers as whole numbers.

1.   A club has 42 members. If $2/3$ of them were present at a meeting, how many people attended? _____

2.   Mort gave away $4/9$ of his 36 comic books. How many did he give away? _____

3.   If $1/16$ of the 96 Americans in the Olympics won medals, how many winners did the U. S. have? _____

TOPIC 8   PHASE II   Supplementary Sheet I

8-1-3-I   Find ALL the replacements for each n which make the given fractions equivalent.

$\dfrac{4}{7} = \dfrac{n}{21}$ _____ $\dfrac{7}{9} = \dfrac{21}{n}$ _____

$\dfrac{11}{17} = \dfrac{n}{34}$ _____ $\dfrac{6}{7} = \dfrac{78}{n}$ _____

$\dfrac{0}{21} = \dfrac{0}{n}$ _____ $\dfrac{7}{7} = \dfrac{n}{n}$ _____

$\dfrac{2}{n} = \dfrac{n}{8}$ _____ $\dfrac{2}{n} = \dfrac{n}{32}$ _____

TOPIC 8   TEST 2-A   Name_____

8-2-2-A   Solve the following problems.
           1. The distance from Ajax to Bathole is 140
              miles. Sam started walking at Ajax and walked
              1/5 of the distance the first day, 1/7 of the
              remaining distance the second day, and 5/16
              of the remaining distance the third day. How
              many miles does he have to go to Bathole at
              the end of the third day?
           2. Joe has 56 record albums. Saturday morning
              he played 3/7 of them. Saturday afternoon he
              played 3/8 of them. How many more did he
              play in the morning than in the afternoon?

8-3-1-A   Find the following products in simplest form:
           1. $\frac{9}{16} \times \frac{7}{10} =$ 2. $\frac{33}{100} \times \frac{90}{99} =$

           3. $\frac{13}{15} \times \frac{10}{99} =$ 4. $\frac{4}{5} \times \frac{17}{17} =$

           5. $\frac{13}{2} \times \frac{10}{7} \times \frac{28}{26} =$

8-4-3-A   Write the simplest form for each fraction:
           1. $\frac{130}{245} =$ 2. $\frac{68}{102} =$ 3. $\frac{216}{396} =$

           4. $\frac{624}{813} =$ 5. $\frac{3108}{10101}$

        Student variables measured in the study included 10
(California Test of Mental Maturity), computational skills
(arithmetic subtest of the California Achievement Test),
knowledge of mathematical concepts (SMSG Mathematics Inventory), attitudes toward mathematics (Aiken and Dreger's
Opinionnaire), test anxiety (Test Anxiety Scale for Children),
flexibility of intra-class group membership, and reactions
toward the experimental procedures.

STUDENT QUESTIONNAIRE

Name_____   Date_____

The following are opinions which might be expressed by

students in this course.   For each opinion, circle whether
you agree (A), disagree (D), or are undecided (U).

Objectives

A   U   D   1.   The objectives and examples are easy to un-
                 derstand.
A   U   D   2.   My summary sheet after the first test on
                 each topic shows me what I have learned and
                 what I still have to learn.
A   U   D   3.   My objectives help me learn when I am
                 studying on my own.
A   U   D   4.   It is not necessary to study the objectives in
                 order to learn the math in this course.
A   U   D   5.   The list of objectives gives me an incentive
                 to learn the math in this course.
A   U   D   6.   Having the list of objectives really puts the
                 pressure on you to learn.
A   U   D   7.   The long list of objectives makes this course
                 seem hard.
A   U   D   8.   I feel discouraged by the number of objectives
                 I am expected to achieve.
A   U   D   9.   I feel discouraged when I think of the number
                 of objectives I have failed to achieve.
A   U   D   10. I have a feeling of satisfaction or accomplish-
                 ment when I look at my list of achieved ob-
                 jectives.
A   U   D   11. My general feeling toward the objectives is
                 a positive feeling.

If you have worked on A level objectives, answer the following
                            questions.
A   U   D   12. The A level objectives are not very interest-
                 ing.
A   U   D   13. The A level objectives provide me with a
                 challenge.
            14. Put an X in front of the choice with which
                 you agree.
                 ____I prefer working on B and I level objec-
                 tives.
                 ____I prefer working on A level objectives.
                 ____I have equal preference for all types of
                 objectives.
A   U   D   15. I do not like being in the A group because
                 then you have more work to do.

Tests

A U D 16. The objectives clearly indicate what kinds of
              questions will be on the tests.
A U D 17. The objectives are <u>not</u> much value in prepar-
              ing for a test.
A U D 18. The tests don't really seem to be testing the
              objectives.
A U D 19. It is very helpful to know just what kinds of
              questions will be on each test.
A U D 20. Having a test question on every objective
              makes the testing fair.
A U D 21. On each test I would rather have fewer ques-
              tions which just sample the objectives even
              though I would then not know exactly which
              objectives would be tested.
A U D 22. The testing lets me know exactly what I
              have learned and what I still have to learn.
A U D 23. The first test on each topic usually puts me
              in group (B, I, or A) in which I feel I
              should be.
A U D 24. The tests are given for only one reason - so
              that my teacher will know what I do and do
              not know.
A U D 25. The tests are useful to me as well as to my
              teacher.
A U D 26. The tests help me in learning the mathemat-
              ics in this course.
A U D 27. So much time seems to be spent in prepar-
              ing for tests that there does not seem to be
              time for the mathematics in the course to
              really "sink-in."
A U D 28. The tests make for competition between stu-
              dents.

RESULTS OF STUDENT QUESTIONNAIRE

Table 37 gives the results of the Student Questionnaire
for the three times it was administered. The entries in the
table are the percent of students responding in each of the
possible categories determined by the items. Significant
percentages are indicated only for the responses of the total
class.

TABLE 37

RESPONSES TO ITEMS ON STUDENT QUESTIONNAIRE

Item		November			January			May		
		A	U	D	A	U	D	A	U	D
1.	Boys	67	5	28	67	33	0	86	7	7
	Girls	64	27	9	76	14	10	76	14	10
	Total	66	17	17	72**	22	6	81**	11	8
2.	Boys	83	17	0	80	13	7	80	7	13
	Girls	86	5	9	76	19	5	90	10	0
	Total	85**	10	5	78**	16	6	86**	9	5
3.	Boys	61	28	11	47	40	13	73	7	20
	Girls	73	4	23	62	14	24	90	0	10
	Total	68*	15	17	56	25	19	83**	3	14
4.	Boys	6	33	61	27	53	20	20	20	60
	Girls	18	23	59	29	52	19	25	15	60
	Total	12	28	60	28	53	19	23	17	60
5.	Boys	28	39	33	27	46	27	27	40	33
	Girls	29	47	24	42	29	29	75	10	15
	Total	28	44	28	36	36	28	54	23	23
6.	Boys	17	11	72	7	27	66	13	0	87
	Girls	5	27	68	9	29	62	5	5	90
	Total	10	20	70*	8	28	64	8	3	89**
7.	Boys	33	11	56	33	7	60	20	0	80
	Girls	18	41	41	24	14	62	0	5	95
	Total	25	28	47	28	11	61	8	3	89**
8.	Boys	0	22	78	7	60	33	13	7	80
	Girls	0	29	71	10	33	57	14	5	81
	Total	0	26	74**	9	44	47	14	6	80**
9.	Boys	6	22	72	13	47	40	47	6	47
	Girls	15	40	45	24	24	52	38	19	43
	Total	11	32	57	20	33	47	42	14	44

*Indicates significance at .05 level of probability.
**Indicates significance at .01 level of probability.

TABLE 37 (cont.)

Item		November			January			May		
		A	U	D	A	U	D	A	U	D
10.	Boys	61	17	22	47	40	13	80	13	7
	Girls	76	24	0	52	43	5	62	14	24
	Total	69*	21	10	50	42	8	70*	14	16
11.	Boys	--	--	--	--	--	--	60	27	13
	Girls	--	--	--	--	--	--	72	9	19
	Total	--	--	--	--	--	--	66*	17	17
12.	Boys	15	23	62	0	56	44	15	0	85
	Girls	31	31	38	30	30	40	8	23	69
	Total	24	28	52	16	42	42	11	11	78**
13.	Boys	46	38	16	22	56	22	85	15	0
	Girls	38	46	16	60	30	10	77	15	8
	Total	42	42	16	42	42	16	80**	15	5
14.	Boys	8	23	69	12	44	44	23	54	23
	Girls	15	31	54	10	40	50	29	57	14
	Total	11	27	62	11	42	47	25	56	19
15.	Boys	--	--	--	0	40	60	0	17	83
	Girls	--	--	--	0	20	80	4	0	96
	Total	--	--	--	0	30	70*	24	7	69*
16.	Boys	50	39	11	75	19	6	80	13	7
	Girls	64	32	4	67	19	14	76	5	19
	Total	57	35	8	70*	19	11	78**	8	14
17.	Boys	0	33	67	6	31	63	7	7	86
	Girls	5	19	76	10	33	57	0	10	90
	Total	3	25	72**	8	32	60	3	8	89**
18.	Boys	11	22	67	19	56	25	7	7	86
	Girls	4	23	73	19	19	62	0	0	100
	Total	8	22	70*	19	35	46	3	3	94**
19.	Boys	83	6	11	50	43	7	87	13	0
	Girls	86	14	0	81	19	0	100	0	0
	Total	85**	10	5	68*	30	2	94**	6	0

TABLE 37 (cont.)

Item		November A	U	D	January A	U	D	May A	U	D
20.	Boys	61	22	17	50	31	19	67	33	0
	Girls	50	50	0	52	43	5	71	29	0
	Total	55	37	8	51	38	11	69*	31	0
21.	Boys	6	33	61	6	56	38	7	40	53
	Girls	4	54	42	10	50	40	5	29	66
	Total	5	45	50	8	53	39	6	33	61
22.	Boys	72	22	6	50	38	12	80	7	13
	Girls	82	18	0	67	24	9	95	5	0
	Total	78**	20	2	59	30	11	88**	6	6
23.	Boys	43	25	32	62	0	38	67	26	7
	Girls	52	43	5	72	14	14	75	20	5
	Total	49	35	16	68*	8	24	71*	23	6
24.	Boys	--	--	--	19	38	43	27	13	60
	Girls	--	--	--	35	45	20	35	15	50
	Total	--	--	--	28	42	30	32	14	54
25.	Boys	--	--	--	69	25	6	86	7	7
	Girls	--	--	--	52	29	19	70	5	25
	Total	--	--	--	59	27	14	79**	6	17
26.	Boys	--	--	--	62	25	13	72	14	14
	Girls	--	--	--	62	33	5	90	5	5
	Total	--	--	--	62	30	8	82**	9	9
27.	Boys	--	--	--	--	--	--	7	13	80
	Girls	--	--	--	--	--	--	20	10	70
	Total	--	--	--	--	--	--	14	11	74**
28.	Boys	--	--	--	38	50	12	54	13	33
	Girls	--	--	--	9	48	43	20	10	70
	Total	--	--	--	22	48	30	34	12	54

TEACHER LOG SHEET

CHAPTER_____   TOPIC_____ PHASE_____   DATE___

ACTIVITY	PLANNED	ACTUAL
1. Whole class time		
2. Group B time		
Group I time		
Group A time		
3. Individual time		
4. Testing		

COMMENTS
1.   Computers and programmed materials
2.   Problems
3.   Miscellaneous

REPORT TO PARENTS

Mathematics 7

Name_____
Home Room Teacher

Date_____

The material for seventh grade mathematics has been written as a detailed set of behavioral objectives. For each topic, the students are grouped three ways on the basis of their initial achievement of the objectives for the topic, and then are given further instruction at the level corresponding to the group to which they belong.
The three levels of objectives and groups are:

> Basic: the minimum that is expected of the students;
> Intermediate: more difficult than basic and about what is expected of average students;
> Advanced: more difficult than intermediate, together with additional content at a more advanced level.

(The groups are referred to as B, I, and A.)

Below are listed the topics covered during the fourth quarter, the group assignment for each topic, and the percent of the group objectives achieved by the end of the topic.

Topic 11: Ratio and Proportion
    General objectives:
        1. To study the concepts of ratio and proportion.
        2. To learn methods for solving everyday problems using ratio and proportions.
                Group_____ Objectives achieved____

Topic 12: Per Cent
    General objectives:
        1. To review the concept of a per cent.
        2. To learn to solve the three types of per cent problems.
        3. To solve word problems dealing with commission, interest and discount.
                Group_____ Objectives achieved____

Topic 13: Statistics and Graphs
    General objectives:
        1. To become familiar with the vocabulary of elementary statistics.
        2. To compute simple statistics.
        3. To construct and interpret graphs of data.
        4. To conduct and report on a sampling experiment.
                Group_____ Objectives achieved____

Comments:

Findings

1. The treatment students showed significantly greater flexibility in their group membership ($p < .01$) than did the control students.

2. A linear regression analysis yielded significant multiple correlation coefficients ($p < .01$) indicating that IQ, computational skills, knowledge of mathematical concepts, or attitudes toward mathematics could have been chosen to assign or predict student group membership.

3. Four subgroups of students--homogeneous with respect to IQ, mathematical achievement, attitudes, and group membership--were identified. Each of these subgroups exhibited different interactions with the experimental treatment.

4. Significant gains ($p < .01$) occurred in computational skills and knowledge of mathematical concepts.

5. No significant differences in means were found between the treatment students and four seventh grade comparison groups with respect to knowledge of mathematical concepts.

6. The treatment group showed a significant gain in attitudes toward mathematics ($p < .01$). Test anxiety decreased, but this result was not statistically significant.

7. The mean gain in attitudes toward mathematics of the treatment students was significantly higher ($p < .01$) than that of each of the four comparison groups.

8. Student reactions to all aspects of the classroom management procedure were favorable. Attitudes toward the procedure at the end of the year were more favorable than at the beginning of the year.

9. The classroom management procedure underwent modification in light of experience with its use. The changes were designed to increase the effectiveness of some procedures and to reduce the disadvantages of others.

Discussion

Although the preparation of objectives and materials to accompany them was time consuming, the classroom use of behavioral objectives offered advantages to the students and to the teacher. The objectives guided the selection of appropriate conditions of learning, facilitated communication between the teacher and students of exactly what the students were to learn, and provided a basis for assessing and differentiating student achievement. To improve this classroom technique, further work should be done in preparing objectives for use in teaching problem solving, discovery, generalization, and proof.

While the classroom management procedure was successful in providing for the individual needs of various types of students, modifications which produce greater achievement during the initial teaching phase of a topic should be investigated. The success of the classroom management procedure with seventh-grade students suggests that this technique should be investigated at other grade levels and in other mathematics courses.

The above discussion summarizing the findings and conclusions obtained from the investigation of the three major parts of the study has indicated the possibility of using intra-class grouping based on achievement of behavioral objectives as an effective form of classroom management. Although this development did result in a workable plan, further modifications should be made and research conducted in order to make this method more suitable for general use in a mathematics classroom.

The first modification would involve having most of the material, especially the objectives, prepared beforehand. Since the continual writing of objectives was found to be a time-consuming process, it would be more efficient and effective to have these materials available to the teacher at the beginning of the course. The format of these materials could take the following possible forms: objectives that cover a wide range of content suitable to a given level of mathematics; objectives that complement an already existing textbook, i. e., type of objectives used in the study; or objectives that are included in the form of the major portion of the text material used in the classroom.

The danger of using any of these types of prepared materials is that they may result in a "canned curriculum." As a sidelight of the present study, it was found that the teacher has to be personally involved in the selection of objectives if he is going to use them effectively. Without personal involvement, it would be easy for the objectives to be used slavishly or not at all.

Ideally, the individual teacher should be the person to write his own objectives, probably with the assistance of a resource person. Another possibility would be to have a team of teachers responsible for a given level of mathematics prepare the objectives for their course. However, the demands of time make these methods of writing objectives impractical in most situations. As one possible way in which the teacher could be spared the initial investment of time in writing the objectives and still be personally involved in the actual selection of objectives for his course, the format of prepared materials might specify the content covered by an objective and then give a list of possible behaviors suitable for this content. The teacher would then be able to select the objectives he wished to use. For a given objective, he would be able to choose the behavior and assign the levels appropriate for his particular style of teaching and the type of students being taught. In this way, the teacher would have the final decision about the structure of his course.

Another area that needs modification, if not initial work, is the use of objectives for instruction in the processes of mathematics: generalization, problem solving, discovery, and proof. In the present study, objectives were not used in these areas. The teacher relied on the progression of the content of the course and the types of exercises given in the textbook for some initial development at work in problem solving and proof. One difficulty in using objectives for this development lies in the fact that these processes do not readily lend themselves to definable overt behaviors. While it is sometimes apparent what the terminal behavior should be, it is difficult to identify the necessary subordinate behaviors, or to assign levels of proficiency to these behaviors.

Another difficulty associated with the use of process objectives in the classroom management plan under consideration is in what phase of a topic the teaching of these objectives should take place. If it is done during the initial teaching phase, one encounters the problem discussed above of making the behaviors, and thus the instruction, meaningful and relevant to all of the students in the class. If the instruction is carried out during the second phase of a topic when the students are divided into groups, there is a problem of fluctuating group membership which makes long range development of these processes difficult.

One method for developing the behaviors associated

with process objectives would be to isolate these behaviors into short separate topics and maintain stable groups during the topics over the course of the year.   Another method would give everyone the initial content required and then, by means of a sequence of increasingly difficult examples and exercises, allow the individual student to move as far as he wants to or can go.   These two methods would have to be highly structured, more so than other management techniques involved in the general plan.   A third method would give all students a goal and the basic outlines of a variety of procedures that can be used in attaining the goal. The students would be allowed to choose the procedure they want to use.

The three methods given in the preceding paragraph are only general outlines of possible strategies that could be used in teaching non-content objectives.   While none of these procedures has been tried within the general framework of the classroom management plan, some such procedure should be considered as an important and necessary modification of this plan for general use.

A final possible modification in the area of content and objectives concerns the materials for the Advanced (A) group.   The study pointed up the necessity for having a wide variety of materials suited to the mathematical abilities of these students.   In addition, the results of the study led to the conclusion that the activities in the A-group should be structured to meet the abilities of two types of students: those who show high performance only because of their inter-action with certain topics and those who show continual high performance because of high general ability.   This modifica-tion in this area should include the development of both short range and long term objectives designed for these two groups of students.

The results of the study indicate the possibility of im-provement in a major area related to the use of flexible grouping.   No provisions were made for handling individual differences or difficulties pertaining to the content covered during the initial teaching phase of a topic.   It was felt that performance, and therefore group membership, could have been improved if the teacher had had some initial indication of how the instruction was being received by the students during this phase.   This was especially true for those stu-dents whose first test placed them at the upper boundary of performance for either the B- or I-group.   If the difficulties of these students had been known to some degree before the first test, it might be expected that a minimum amount of additional instruction would have placed these students in a higher group which would more accurately reflect their needs during the second phase of the topic.

As a possible modification of Phase I which would incorporate this early diagnosis and yet conform to the necessities of pacing involved, it is suggested that some form of frequent--if not daily--evaluation be used in the initial teaching. This evaluation could take the form of short quizzes given either immediately following a lesson or at the beginning of the next class period. By focussing on the behaviors associated with the major objectives being studied, the two or three items contained in these quizzes would provide the teacher with immediate feedback about relevant student performance. The teacher could respond to low performance by restructuring his teaching of particular objectives for the whole class or providing small group or individual instruction of other objectives. It is felt that this method of continuous evaluation could be inserted in the procedures for Phase I without drastically increasing the amount of time allocated to this phase of a topic. The benefits that would hopefully be derived from this modification, in terms of increased achievement of objectives, seem to justify any additional time required for its use.

The discussion of modifications designed to improve the effectiveness of the method of classroom management explored in this study has implicitly identified areas for possible further research. In general this research should involve exploration of the effectiveness of different combinations of the original and modified aspects of the experimental treatment. In addition to these explorations, the following two areas are also suggested for further research.

1. Verification studies of the use of behavior objectives and intra-class grouping should be conducted. The present study focussed on initial exploration and development of the combined use of these techniques. The findings of the study indicated the need for controlled comparisons of this method with other methods of classroom management used in mathematics. The results of such studies would indicate more definitely whether or not the time and effort expended by using the experimental treatment in the classroom was worthwhile in terms of the broad goals of mathematics education.

2. Studies should be designed involving students at other grade levels. The present study used two classes of seventh-grade students. Most other studies of intra-class grouping have been confined to the elementary school. Studies of the use of behavioral objectives in the classroom are virtually non-existent. In order to ascertain the effectiveness of these techniques for general use, results are needed from a wide range of age groups and mathematics courses.

The results of this study and the suggestions for further modifications and research do not give definite answers to the problems of effective provision for individual differences and efficient management procedures for accommodating these provisions in the mathematics classroom. However, they do show that the development of methods built around flexible intra-class grouping and behavioral objectives has the potential of meeting these important goals of mathematics education.

COMPUTER-ASSISTED INSTRUCTION OF
UNDERACHIEVING CULTURALLY DEPRIVED STUDENTS

Marian J. Patterson

Many articles have been written about the learning experiences of average students who have been exposed to computer-assisted instruction. This article, however, is concerned with underachieving culturally disadvantaged Mexican-American or Black students.

Two seventh-grade classes were given The Iowa Test of Basic Skills and The Wide Range Achievement Test. On the basis of these test results 20 students were selected by the teacher-coordinator to participate in computer-assisted instruction for a period of 34 days during the fall semester of the academic year 1968-1969.

The test scores were used by Stanford Computation Center to place each student in a "class" and to program him to receive drill-and-practice lessons containing certain concept blocks. The 20 students were grouped into three "classes." Twelve concept blocks were planned for these 20 students.

The three "classes" were third grade, fifth grade, and sixth grade. Two students made up the third-grade "class," 12 the fifth-grade "class," and six the sixth-grade "class."

The students were programmed to study the following concepts (not necessarily in the order given):

Third Grade

Mixed drill of whole numbers fractions, addition, subtraction and reducing to lowest terms
Multiplication of one and two digit numbers
Addition and subtraction of whole numbers (easy)
Addition and subtraction of whole numbers (harder)

Fifth Grade

Mixed drill, addition and subtraction
Multiplication of two and three digit numbers
Fractions, addition, subtraction, reducing to lowest terms
Division, finding partial quotients only
Mixed drill, including fractions and decimals

Sixth Grade

Mixed drill of whole numbers
Fractions, addition, subtraction, multiplication and division
Division, finding partial quotients only
Mixed drill, including fractions and decimals
Mixed drill, including fractions, division, per cent

The drill-and-practice lessons were written by Dr. Patrick Suppes of Stanford University and his associates. The students received their lessons on a teletypewriter terminal which was connected by telephone to a computer located at the Stanford Computation Center.

There are five levels of difficulty for each lesson within a concept block. These drill-and-practice lessons were automatically presented, evaluated, and scored by the computer.

The 20 students shared one teletypewriter. The teacher-coordinator scheduled the students so that they could use
the teletypewriter at least once a day during a two hour interval. Occasionally students were able to take two lessons
in one day.

A new concept or idea was introduced by the teacher-
coordinator prior to the student receiving this lesson on the
teletypewriter. The concept was usually taught during the
preceding day.

After being programmed into a "class" and a concept
block the student's first work at the teletypewriter consisted
of a pre-test. The computer then branched to one of five
levels based on his performance. Students who scored between 60 and 79 per cent were given another lesson on the
same level the following day; those who scored above 79 per
cent were given a lesson on the next higher level; and those
who failed to score at least 60 per cent were given a simpler
lesson on a lower level. After the student had completed a
concept block, the machine administered a post-test. The
results of this post-test were used by the computer in selecting the next concept block or in selecting review material
for the student. Thus the computer provided highly individualized review and practice lessons on basic concepts and
skills.

A time interval of ten seconds was given for a response. If the student gave the incorrect response, then the
exercise was retyped. Whether the student gave the incorrect response or took too much time, he was always given a
second chance. The students were not allowed to use pencil
or paper. Each exercise was printed by the machine in a
format that required only a single response. The response
was limited to either numerical answers or simple single-
letter answers for multiple-choice exercises.

The machine was successful as an attention-getting
device, in eliciting responses, in keeping students seated
during the lesson, and in cementing concepts and ideas
through drill. The teacher-coordinator of these culturally
deprived, underachieving students appreciated the receiving
of help in those areas.

Parents and students evinced interest and enthusiasm

during the entire program.   Attendance was good even though
the lessons started one hour before the opening of school.

INDIVIDUALIZING ARITHMETIC-TEACHING

Eugene R. Keffer

Evidence available from general aptitude, special aptitude, and achievement tests and observations by highly
skilled teachers of arithmetic leads us to conclude that not
all pupils have the ability to master arithmetic skills according to a set schedule.   And yet, quite frequently all pupils
in any given class are taught the same arithmetic principles
and processes simultaneously.

While it is not our purpose to minimize the importance
of a pupil's needs and goals as conditioners of readiness, in
this article we are more concerned with aptitude and achievement as readiness factors in learning.   Maturation, the natural growth of the physical bases for mental functions, conditions the ability to learn.   Where there is insufficient maturation, it may be wholly impossible for a child to utilize the
related mental functions.   Mental development, the progressive growth and organization of the mental functions and
psychological behavior of the individual is conditioned by both
maturation and learning.   Consequently, former experiences,
maturation of the somatic bases of learning, and previous
exercise of functions determine, to a large extent, the degree
of readiness which a pupil possesses.

Experiments in which standardized psychometric devices have been used to gather the data have demonstrated to
us that there are major differences among the pupils in the
typical classroom of an American school.   Among the differences which have been isolated are the following: mental
age, level of achievement, and specific aptitudes.

In this paper, we propose a method of teaching arithmetic which takes cognizance of the aforementioned factors
and which is organized in such a manner that each pupil is
encouraged to develop arithmetic skills in accordance with
his readiness and application.   This method has been used
very successfully for many years in the sixth grade of the

College Laboratory School of Central Missouri State College.

In this teaching method, the classroom is organized as a laboratory.   During the first week of school, the teacher attempts to ascertain the level of achievement of each pupil in the class.   The criteria which are used to ascertain the level of achievement are: teacher observation of each individual pupil's work, teacher-made and textbook tests, and conferences with each pupil to determine the highest achievement level at which he understands the arithmetic principles and processes.   Each pupil, then, is given his initial assignment at the achievement level where he can work with understanding.

Before assignments are made pupils are urged to calculate problems which they can work independently.   Textbooks at the fifth-, sixth-, and seventh-grade levels are made available to them   Enrichment materials are provided in order to keep all pupils busy during this period of pupil classification.

After all pupils are classified, instruction and assignments are given.   Each individual at a specific level of achievement is invited to the desk of the teacher who explains and illustrates the principles and procedures necessary to solve problems at this level.   Then the pupil works a problem on the blackboard to demonstrate his knowledge and understanding of principles and procedures involved.   When the individual has demonstrated that he knows and understands these principles and procedures, he is given an assignment which does not go beyond the principles and procedures he has learned.

If two or more individuals are at the same level of attainment at any given time, the teacher instructs them, at her desk, in the same manner that she used with the individual pupil.   Each individual in the group must demonstrate his knowledge and understanding of the principles and processes involved before any assignment is given to him.

During the regular class period, following the assignment, the learner works the problems at a speed which he chooses until he completes the assignment or until he doesn't understand what he is doing.   He keeps a notebook in which the problems are organized so that the teacher can spot-check them.   From this spot-checking the teacher determines his progress, his understanding of the problems and the areas where more understandings need to be developed.

When the teacher is not explaining a new process during the class period, she circulates among the pupils and checks their notebooks to determine the degree of mastery and understanding which they have attained relative to the problems they are working. If they need drill exercises in order to develop mastery and/or understanding of the problems, these exercises are assigned; otherwise, no drill exercises are assigned. If review or relearning is indicated, individual instruction is arranged. Thus, valuable time is conserved which is used to relearn processes and/or principles which were forgotten or to learn new principles and processes when mastery has been achieved on a given assignment.

Occasionally the entire class may be taught the same principle(s) and/or process(es) simultaneously. The entire class is taught simultaneously, however, only if all the pupils have the background to understand what is being taught. For example, the pupils may be constructing a map of Italy during their study of the geography of the land, and lack knowledge and understanding of Roman numerals, area, perimeter, and addition and subtraction of measurements. In this instance the teacher would instruct the pupils simultaneously in these areas of deficiency.

The teacher must be very familiar with the arithmetic taught in grades four, five, six, seven, and eight in order to plan for the needs of each individual pupil. She may have to teach pupils the principles and processes which they have failed to learn or have forgotten, and she may find it necessary to teach principles and processes which ordinarily are taught in the seventh or eighth grades.

In order to provide materials for pupils, it is necessary to provide textbooks by several authors and publishers for each of the grade levels indicated by the levels of achievement represented in the class. Although one particular textbook is chosen for each grade level as the text for that grade level, problems in it are not reassigned if mastery is not attained when they are worked the first time. Rather, similar problems in another textbook are assigned for drill and/or relearning.

When instruction is individualized, it is very necessary that evaluation of pupil progress be continuous. In addition to continuous evaluation, each pupil is given a written test in problem-solving and computation at the end of each unit or chapter that he completes.   The teacher checks the procedures and answers and returns the test to the pupil, who is instructed to check the problems and locate his errors if there are any.   If he cannot find them, the teacher identifies them for him.   Then she assists the child to correct the errors.   Where he lacks understanding, the work is repeated until mastery and understanding are attained.   After the pupil masters the principles and processes, he is given individual instruction in a new process and/or principle and is given a new assignment.

In order to control feelings of inferiority or superiority, the teacher discusses with the pupils the various degrees and kinds of abilities which are expected in the classroom. These pupils are taught to accept their limitations and their strengths.   They are taught to be dissatisfied if they don't try to make progress commensurate with their abilities and not to feel inferior or superior if they are above or below the level of attainment of their classmates.   No feelings of inferiority or superiority have been detected by the teacher. When pupils are very close in their assignments, some competition may develop.   The competition seems to stimulate the persons involved to progress more rapidly than they would under other circumstances.

AN EXPERIMENTAL STUDY OF INDIVIDUALIZED INSTRUCTION IN GENERAL MATHEMATICS

George C. Nix

The problem of this study was to analyze two methods of instruction in general mathematics in order to determine their effect on student achievement, student reaction, and teacher effort.   The two methods of instruction used were group-oriented instruction and individualized instruction.

Specifically the purposes of this study were:

1.   To discover under which of these methods of instruction students achieved most in general mathematics.
2.   To discover if students of different age, sex, and

previous levels of achievement achieved more under one of
these methods of instruction than under the other.

    3. To discover student reactions to the method of
individualized instruction in general mathematics.

    4. To discover if the teacher must spend more time
in out-of-class preparation for the individualized method of
instruction than for the group-oriented method.

At the beginning of the school year, six eighth-grade
classes were selected for this study; three classes each for
the control and experimental groups.   All students were
given the Stanford Achievement Test and the Otis Quick Scoring Mental Ability Test.   The classes in the control group
were taught under the group-oriented method of instruction
and the classes in the experimental group were taught under
an individualized method of instruction.   A time log was
maintained on the daily out-of-class preparation time of the
teacher.   Near the end of the school year all students were
given the mathematics portion of the Stanford Achievement
Test as well as a questionnaire on student reactions to the
method of instruction.

The control group and the experimental group were
divided into sub-groups according to sex, age, intelligence,
and over-all achievement and mathematics achievement,
based on test data obtained in September.   The pre-test and
post-test of the mathematics portion of the Stanford Achievement Test were used to determine the change in mathematics
level for each student.

In order to achieve purposes one and two of this
study, the mean mathematics achievement level change of
each sub-group was computed.   Then the means of each pair
of corresponding sub-groups were subjected to the t test for
significant differences between two means.   The third purpose
of this study was achieved by analysis of the student questionnaire.   The fourth purpose of this study was achieved by
analysis of the teacher's time log on out-of-class preparation
for each of the two methods of instruction.

Definition of Terms

GROUP-ORIENTED METHOD: This term implies a
method of instruction used by the teacher in which the class
is taught as a group.   The teacher lectures to the group as
a body.   Each group member has the same textbook, gets

the same assignments, takes the same tests at the same
time, and goes on to the next unit for study with the group
regardless of whether the present unit is sufficiently understood.

INDIVIDUALIZED METHOD: this term implies a
method of instruction used by the teacher in which each member of the class is taught individually or in small groups according to his or her needs.   The group is not lectured to
as a body.   Each member works at his own speed, at his
own achievement level, in a textbook based on his own
achievement level, takes a test on a unit only when he feels
confident he can pass the test, and advances to the next unit
for study only when he has mastered the present unit.

GENERAL MATHEMATICS: this term refers to mathematics courses that involve the elements of arithmetic.
These mathematics courses may include fundamentals, but
not full courses, of algebra and geometry.   These are the
mathematics courses usually given in the upper elementary
and junior high grades.   At times in the study, the term
"mathematics" is substituted for "general mathematics. "

Control Group

Group-oriented instruction was the instructional method
used with the control group.   An effort was made to instruct
the students in these sections in a matter that was traditional
and as near as possible to the method of instruction that the
students had been accustomed to in the past in their mathematics classes.   To determine the procedures to be used in
the group-oriented classes, conferences were held with several mathematics teachers and elementary school teachers who
taught mathematics in their grades.   This was done in order
to benefit from the practices of these experienced teachers.

The students who were taught under the group-oriented
method of instruction were all issued the same eighth-grade
level textbooks.   They all had the same homework assignments, took the same examinations at the same time, and
were expected to work at the same rate of speed.

The students reported to class, the roll was checked,
and homework collected.   The teacher usually started the
class by demonstrating the procedure from the homework by
working a few examples on the chalkboard.   The teacher

often asked the students questions about the procedures pertaining to problems in the assignment, and the students would
also ask the teacher to explain how to work certain problems.
At times the teacher would ask one or several students to
work problems from the homework assignment on the board.
These students would then explain their work to the rest of
the class. Generally the teacher did most of the talking and
the students listened. This was the lecture-demonstration
type of teaching and definitely was group-oriented.

If the class period ended before the necessary explanation of the procedures were completed, then the class would
take up at that point the next day. However, a few problems
pertaining to those procedures were given as homework.

When the teacher felt that all of the material of the
old assignment had been finished, the entire class moved on
to the pages in the textbook that contained the new material
and procedures to be taken up next. As with the old exercise, the teacher would go over procedures for the new
exercise and make an assignment. This assignment was
usually in the form of problems to be worked on paper and
handed in the next day. Any time left in the class period
was spent in supervised study of procedures for the new
exercise and in getting the class started on the homework
assignment. The next day the same methods were followed.

When the class completed all the exercises in a unit,
or spent an alloted number of days on the unit, the teacher
usually reviewed the procedures in the unit. Every one in
the class took the same examination at the same time. Although some of the students would not pass the examination,
the entire section would go on to the next unit together.
The same general procedures or methods were followed unit
after unit throughout the entire textbook or until the end of
the school year.

Experimental Group

Individualized instruction was the instructional method
used with the experimental group. In these sections the instruction was to one student at a time, or to small groups,
rather than to the class as a whole. Each student received
individual help from the teacher or from other members of
the class as needed. However, at times the teacher would
talk with the class as a group, when it seemed that most of

the class were having trouble with similar concepts.   These
times were infrequent and brief.

Each student was issued a textbook according to the
grade level he had mastered as determined by his score on
the mathematics achievement test.   No one in the study was
given a textbook below sixth-grade level nor above eighth-
grade level at the start of the study, but the student was
moved to a lower level book if he could not master the one
given him, or to a higher level book upon mastering the one
given him.   No programmed materials or teaching machines
were used.   The instructor made available only the materials
that would normally be found in the typical eighth-grade
mathematics classroom.   The classroom contained the same
supplementary texts and mathematical aids that were available to students in the control group.

The students reported to the class, the roll was
checked, and each student began the exercise that he needed
to work at that day from his own textbook.   Each student
studied each unit in his textbook at his own rate of speed
which was dependent upon his understanding the material in
the unit.   When he felt that he was ready to be tested in the
unit he was given such a test.   If he passed the test he went
on to the next unit.   If he did not make the required score
to pass, then he had to go over the unit again.   This time
the teacher required that he hand in certain problems from
the unit, depending on what he had missed on the test.   The
teacher checked this work to help the student locate his
weaknesses.   Problems missed on the test were discussed
with the student whether or not he made a passing score.
After this required work was handed in and checked, the
student, when he felt that he was ready for it, took a test
similar to the one taken before.   If he failed it again he
was required to go back to a unit in a lower grade level
textbook, that presented the procedures that were holding
him back.   When he mastered this unit, and passed a test
on it, he returned to the unit in the higher grade level textbook which had stopped him.   He was still required to study,
take and pass another test on this unit that had stopped him,
before he could go on to the next unit in his textbook.

When the student had finished all units in his textbook
he was given the next higher grade level textbook.   When a
student finished the eighth-grade level textbook he was given
a first-year algebra book that introduced elements of algebra,
geometry, and trigonometry with several units that introduced

modern terminology and concepts that were used in algebra
by applying these in general mathematics.

Much of the work of the teacher, of necessity, had to
be done after the end of the typical school day.   The teacher
graded any tests taken that day.   On those that were failing
or near failing the page numbers in the unit that corresponded to the problems missed on the test were recorded on
the test paper.   Those papers handed in that day as homework were looked over for procedures more than for accuracy.   If the procedures were wrong, a note was made on the
paper for the student to do them over after receiving help
from the teacher or from someone in the class who had already mastered those procedures.

A file card was kept on each student.   Each day the
test results, assignments, and dates were recorded.   Copies
of each test were kept on file and a test was pulled for each
student who was ready for testing the next day.   It was necessary that several different versions of each unit test be
made.   This was done early in the year because of the high
achievers and those who wanted to be tested when they were
not actually ready and would often have to be retested.   A
different version of the test was given each time the student
was retested.

The following day, after the roll was checked, homework papers that had wrong procedures were returned.   Those
students who were to be tested were called to a table or
desks near the front or in one corner of the room.   Usually
there were no more than three or four students taking tests
that day.   An effort was made to keep the number of students
being tested to no more than six because it took much of the
class time to go over these tests the next day with the students individually.

The students who were tested the day before were
called upon to go over their tests, one at the time.   If the
student had a high passing score, those problems he missed
were quickly gone over and his next unit assignment was
made.   If the student made a low but passing score, the
problems missed were discussed and those pages in the unit
that covered the problems missed were given to the student
to work.   Upon completion of these problems the student
moved on to his next unit to study.   If the student did not
pass the test, those problems missed were discussed and a
list of pages that covered the problems missed were given to

()

the student to work and hand in.   He was to review all of
the unit but concentrate his effort on those pages indicated.
He would then be retested upon completion of those pages
and when he felt confident that he could pass the test.

After going over the examination papers with the students tested the day before, which usually took about half of
the class period, the teacher moved about the room helping
individual students, especially those whose homework procedures had been wrong on the papers returned that day.   He
got students working together that were on the same unit and
those that needed similar help.   He would explain certain
methods and concepts to these small groups that needed similar help.   He worked with the faster students as well as
with the slower students.

Procedure

A questionnaire was devised for the third purpose of
this study.   This questionnaire, given near the end of the
school year, was for the purpose of determining which of the
two methods of instruction the students preferred and why.
Responses to all items in the questionnaire were analyzed
separately and collectively to give insight into the reasons
given by the students for preferring one method of instruction
over the other method of instruction.

QUESTIONNAIRE

1. Did you like____, dislike____, or feel indifferent____
    about school last year?
2. Did you like____, dislike____, or feel indifferent____
    about school this year?
3. Did you like____, dislike____, or feel indifferent____
    about mathematics last year?
4. Did you like____, dislike____, or feel indifferent____
    about mathematics this year?
5. Do you want to finish high school? Yes____, No____,
    Don't care____.
6. Do you want to take mathematics courses in high school?
    Yes____, No____, Don't care____.
7. In comparing school this year with school last year, do
    you think you learned more____, less____, or about the
    same____ as you did last year?
8. In comparing mathematics this year with mathematics
    last year, do you think you learned more____, less____,

        or about the same____ as you did last year?
   9.   Were your grades this year, in all subjects combined,
        higher___, lower___, or about the same___ as they
        were last year?
  10.   Were your grades this year in mathematics, higher___,
        lower___, or about the same___ as they were last
        year?
  11.   Did you like your classmates this year better___,
        less___, or about the same___ as your classmates
        last year?
  12.   Did you like your teachers this year better___,
        less___, or about the same___ as your teachers last
        year?
  13.   Did the method of instruction used by your mathematics
        teacher this year differ greatly from what you have
        been accustomed to in mathematics? Yes___, No___.
  14.   Would you like to continue under the present method of
        instruction in your mathematics classes? Yes___,
        No___. Why___.

      The fourth purpose of this study was achieved by
maintaining a daily log of the teacher's out-of-class preparation time for each group.   This log contained space for recording the time in five different areas of preparation time.
The total amount of time spent by the teacher in out-of-class
preparation time, as well as the time spent in each area,
for the control group was compared with that for the experimental group.

Hypotheses Tested

      On the basis of the data obtained in this study the fol-
lowing findings were made:

      (1)   Students who were taught under the individualized
method of instruction in general mathematics did not achieve
significantly more than students who were taught under the
group-oriented method.
      (2)   Students in over-all achievement and at all levels
(high, average, and low) who were taught under the individualized method of instruction in general mathematics, did not
achieve significantly more than students of corresponding
over-all achievement levels who were taught under the group-
oriented method.
      (3)   Students in mathematics achievement in high and
low levels, who were taught under the individualized method
of instruction in general mathematics, did not achieve

significantly more than students of corresponding mathematics achievement levels who were taught under the group-oriented method.

(4) Students in mathematics achievement, average level, who were taught under the individualized method of instruction in general mathematics achieved significantly more than students of average mathematics achievement who were taught under the group-oriented method.

(5) Students of all age levels (below 13 years of age, 13 and 14 years of age, and 14 years of age and older) who were taught under the individualized method of instruction in general mathematics, did not achieve significantly more than students of corresponding age levels who were taught under the group-oriented method.

(6) Students of above average and average intelligence, who were taught under the individualized method of instruction in general mathematics, did not achieve significantly more than students of corresponding levels of intelligence who were taught under the group-oriented method.

(7) Students of below average intelligence who were taught under the individualized method of instruction in general mathematics, achieved significantly more than students of below average intelligence who were taught under the group-oriented method.

(8) Girls who were taught under the individualized method of instruction in general mathematics did not achieve significantly more than girls who were taught under the group-oriented method.

(9) Boys who were taught under the individualized method of instruction in general mathematics achieved significantly more than boys who were taught under the group-oriented method.

Questionnaire Data

(10) Students who were taught under the individualized method of instruction in general mathematics showed a greater change to a more favorable attitude toward school and especially toward mathematics than students who were taught under the group-oriented method.

(11) Students who were taught under the individualized method of instruction in general mathematics more strongly indicated they had learned more than previously in school and in mathematics than students who were taught under the group-oriented method.

(12) Students who were taught under the individualized

method of instruction in general mathematics were better
satisfied with their classmates and teachers than students who
were taught under the group-oriented method.

(13)   Students who were taught under the individualized
method of instruction in general mathematics indicated the
method of instruction used with their group was not similar
to the method of instruction of which they were accustomed.

(14)   Students who were taught under the group-oriented method of instruction in general mathematics indicated
the method of instruction used with their group was similar
to the method of instruction of which they were accustomed
and was indeed group-oriented.

(15)   Students who were taught under the individualized
method of instruction in general mathematics had a stronger
desire to continue with the current method of instruction
than students who were taught under the group-oriented method.

Teacher Log

(16)   The teacher spent 164 per cent more time in
making lesson plans and assignments for students who were
taught general mathematics under the individualized method
of instruction than for students who were taught under the
group-oriented method.

(17)   The teacher spent 25 per cent less time in
checking homework of students who were taught general
mathematics under the individualized method of instruction
than of students who were taught under the group-oriented
method.

(18)   The teacher spent 492 per cent more time in
making exams for students who were taught general mathematics under the individualized method of instruction than for
students who were taught under the group-oriented method.

(19)   The teacher spent 20 per cent more time in
grading exams and averaging grades of students who were
taught general mathematics under the individualized method
of instruction than of students who were taught under the
group-oriented method.

(20)   The teacher spent 84 per cent more time in
general preparation for the individualized method of instruction in general mathematics than for the group-oriented
method.

(21)   In all, the teacher spent 38 per cent more time
in out-of-class preparation for teaching general mathematics

under the individualized method than under the group-oriented method.

Not all students achieved more from individualized instruction in general mathematics than from group-oriented instruction.   Certain students, however, did achieve more in general mathematics when taught under one method of instruction than under another.   Students of average mathematics ability, students of below average IQ, and boys achieved significantly more under individualized instruction than the corresponding sub-groups under group-oriented instruction.

In addition, students developed a more favorable attitude toward school and mathematics, learned more mathematics, and were better satisfied with classmates and teachers when taught general mathematics under the individualized method than under the group-oriented method.   To achieve this improvement, the teacher spent more time in out-of-class preparation for individualized instruction in general mathematics than for group-oriented instruction.

DIAGNOSIS AND REMEDIAL INSTRUCTION IN MATHEMATICS

R. B. Thompson

There are many factors which have made it mandatory that teachers modify their instruction.   The old textbook and recitation method of teaching is no longer accepted.   In general, schools do not now hold up standards of attainment and fail to promote pupils on the basis of their failure to master certain predetermined units of instruction.   All the children of all the people are now being promoted by age in heterogeneous groups.   The trend is that the pupil shall no longer be held responsible for his failure to master the subject at hand, but rather that the teacher shall be held accountable. The incentives of punishment and reward are no longer present.   Children are not retained because of lack of subject content mastery.   Instruction has become child centered rather than subject centered.   We teach children--not facts. However, we must not only teach children but we must surely teach them something.   Accepting the challenge that the school must adjust to the child rather than the child to the

school, conscientious attempts are being made to manage instruction so that each individual pupil may profit most by the
time spent in school.   Teachers of the skill and content subjects probably feel this responsibility more than instructors
in other fields of teaching.

In the effort to formulate a plan of instruction whereby
each individual might progress at the optimum rate of speed,
a four-year experiment was carried on under the direction of
the author at the Training School of the University of Utah.
The experimentation was confined to the fields of arithmetic
and algebra.   Instruction was individualized within the group.
Identical plans of procedure were not followed with each
class each year; however, the class organization which was
generally used is here described.   The usual plan of arranging experimental and control matched groups of pupils was
followed.

In the experimental group, each student worked alone
except for individual aid from the teacher.   However, since
no class recitations were held, the teacher was free the full
hour to help and guide each student.   The procedure for each
individual pupil was as follows: Each student took a diagnostic pre-test to determine whether or not the student needed
work in that particular phase of arithmetic.   If it was found
that the pupil already possessed a mastery of that unit of
work, the next pre-test was taken, and so on until the pupil
found a weakness in his mathematical preparation.   When a
student failed a pre-test, remedial drill work was given and
a final test given over the unit to determine whether or not
mastery had been reached.   Three additional tests equivalent
to the pre-test were available.   Thus, when a student failed
the pre-test, he had the opportunity to drill and test three
more times.   Thus no pupil wasted time working on topics
which he had previously mastered.   Each student progressed
at his own rate of speed and did not wait for the whole class
to do all the work, nor did any pupil leave a topic until this
part of the work was thoroughly mastered.   Of course not
all pupils finished all units of work in the same time, but
the teacher was free to spend all his time helping each individual pupil.

In the control group, the regular textbook lesson assignment recitation method was used.   All students prepared
the same lesson each day.   About one-half of the period was
spent in study and one-half of the period was spent in recitation.   All students attempted to learn the same amount of

material in the same time.

During a 10-week period, an experimental group, which we will call Group A, of 28 seventh-grade pupils advanced on an average 1.4 years in arithmetic, while the control group, which we will call Group B, advanced .40 of one year. During the next 10-week period, the two groups were interchanged and Group B, which now experienced individualized instruction, gained .99 of one year while Group A, which now reverted to the textbook procedures, lost .27 of one year.

We might understand the loss in Group A when we consider that many of these pupils were carried far beyond the average of the entire class. Now when the old lockstep textbook plan of teaching was used, the material failed to challenge their ability and they lost interest as well as some of the skill they had attained. When pupils are carried beyond their normal expectations and the situation is then changed so that a challenge is no longer present, some of the skill is certain to be lost. In order for the pupils in Group A to continue to gain, it would have been necessary to continue to challenge the ability of each individual pupil.

In the next year's experiment a seventh-grade class of 38 pupils made an average gain of 1.4 years in ten weeks. In this case at any rate, a class of 38 pupils made as much progress per pupil as a similar class of 28 pupils had made in the same amount of time.

There is one point concerning the permanency of diagnostic and remedial instruction which should be noted. The 56 seventh-grade students who were in the first experiment took algebra in their eighth year of school receiving no arithmetic instruction during this time. At the end of their year of algebra their average grade level in arithmetic was 9.2. Thus, by the end of their eighth year of school these students had taken one year of algebra and stood at grade level 9.2 in arithmetic.

The third-year experiment in arithmetic was carried on in a somewhat different manner than the experiments of the previous two years. The controlled instructional period lasted one full year rather than ten weeks. It was found that some variety of practice materials was advisable in order to maintain interest. Students were taught how to locate supplementary drill material from various sources.

The procedure of test-drill-test was followed until a satisfactory degree of mastery had been reached.   During the
year the gains of 35 seventh-grade pupils ranged from 1.1
years to 4.0 years with an average gain of 2.6 years.

The 38 pupils who entered the seventh grade of the
Training School were followed for three years until the completion of their course in algebra.   With such exceptional
and consistent results over a period of three years it seems
quite evident that diagnostic and remedial individualized instruction is at least one very effective method of teaching
mathematics.

VI. INDIVIDUALIZING MATHEMATICS INSTRUCTION:
3. OTHER APPROACHES

Without a doubt, the introduction of individually prescribed instruction (IPI) has been the most elaborate and extensive experiment in the field of individualization of instruction since its inception.

Jack R. Fisher

This program originated at the University of Pittsburgh, and the first article in this chapter has been excerpted from a University of Pittsburgh doctoral dissertation
by Jack R. Fisher.   Dr. Fisher began his work on individually prescribed instruction at the University of Pittsburgh in
1961, before the establishment of the Learning Research and
Development Center.   This center has further expanded the
IPI concept into several other content areas in addition to
mathematics.

At the present time, Dr. Fisher is a research fellow
with Research for Better Schools in Philadelphia.   He is an
assistant project director in the Individualized Learning Program of that organization.

In my opinion, the availability of federal funds for
educational research has been remarkably adequate in at
least one aspect.   This refers to the establishment of Regional Education Laboratories in all parts of the United States.
In the short period of their existence these laboratories have
made substantial contributions to educational theory and
practice in a way that, in the case of most of these laboratories, has had a national as well as a regional impact.   It
is most unfortunate that the decision to curtail the operations
of some of these laboratories, because of necessary prior-
ities in the demands on the limited amount of funds available,
may have been interpreted by some to mean that the laboratories had failed to make valuable contributions.

Regional Education Laboratory: Carolinas, Virginia

The first item in this chapter is an abbreviated description of a program developed by the Regional Education
Laboratory for the Carolinas and Virginia called Individualized
Mathematics System (IMS).   Here is an example of a plan of
individualization which falls between those plans that require
major expenditures of time and funds (such as computerized
programs) and those such as were described by Keffer and
Nix in the preceding chapter.   It is a carefully worked out
program, worthy of the reader's careful examination and
consideration.   This seems to be a good place to make a
rather obvious statement: anyone wishing to individualize
mathematics instruction, or for that matter any other subject, can probably find the way most congenial and effective
for him by combining features from a number of the programs described here, rather than adopting all the features
of a single plan.

Bernard R. Wolff

I am not at all certain that all readers will agree with
me that the second article in this chapter is descriptive of
an individualized mathematics program of instruction.   I
hesitated in deciding whether this article should be placed
here or in subsequent chapters dealing with research.

The selection consists of parts of a doctoral thesis
written by Professor Wolff at the University of Oregon in
1968.   The purpose of the study was to determine whether
there was a difference between the way teachers in non-
graded classes and teachers in graded classes regarded and
practiced individualized mathematics instruction.

In order to find the answer to this question, Dr.
Wolff, who is now a professor in the Education Department
of Lewis and Clark College in Portland, Oregon, used the
Q-sort technique.   Much of the excerpt describes this technique.   As those who are familiar with the Q-sort technique
already know, this is a procedure where the ranking by the
subjects of a number of statements (usually placed on cards)
is forced by requiring a certain number of cards to be
placed in each of a number of categories.

In this case, the forced ranking procedure was administered to ten teachers who taught graded classes and ten

teachers who taught in non-graded classrooms. The statements themselves seem to me to be of great interest to the
student of individualized instruction. Any reader who is willing to take the time and trouble to rank the 70 statements,
all of which are found in the excerpt included here, can learn
much about himself and what he really believes about individualization of instruction. I have included the results of
the sorting process only for the ten teachers in graded rooms
and only for the first 55 cards. I did not feel that the inclusion of the complete results would be a useful way to fill
space. If any reader sufficiently interested in this particular
aspect of Dr. Wolff's study wishes to see the complete results, he can, of course, purchase a microfilm of the entire
thesis from University Microfilms at a nominal cost. Microfilm readers are now standard equipment in most libraries,
and all the theses from which excerpts are included in this
book are available in microfilm or photocopy form from the
above source.

I have stressed continually the fact that the whole
process of individualization is one that allows much leeway
and freedom not only to the learner, but also to the teachers. This is beautifully illustrated by running down any of
the columns in Table 1 in Dr. Wolff's article, where it is
immediately apparent that the statement which is of top importance to one teacher is classified by another teacher as
being of minimal significance in describing individualized instruction.

Victor L. Fisher, Jr.

The next article in this chapter is excerpted from the
1966 Indiana University thesis of Dr. Fisher, who is principal of an eight-grade elementary school in Evansville, Indiana, where he employs a variety of aspects of and approaches
to individualization of instruction. In addition, Dr. Fisher
is also an adjunct professor at the University of Evansville,
where he teaches courses in mathematics to elementary
school teachers.

The description of Dr. Fisher's project is quite clear
and to the point. As is the case in a number of other
studies, no significant differences were found in the scores
obtained on standardized mathematics tests between pupils
whose mathematics instruction had been individualized and
those who had been taught in the traditional manner. If

individualization is really a better procedure than the traditional approach, why then don't children subjected to such an
approach do better on standardized tests? I think that the
answer to this is not really very complicated. There are a
number of factors that should be taken into account.

Standardized tests in elementary school mathematics
tend to stress those particular skills that are taught in the
conventional way of teaching, using conventional materials
and being guided by conventional textbooks and conventional
teachers manuals. In individualized instruction, the tendency
is to regard an understanding of processes as more important than a mastery of rote processes and procedures without
necessarily understanding their meaning. Let me give an il-
lustration of what I mean. Here is a problem which was
actually used in a doctoral experiment:

"Mary bought a 3/4 pound box of candy which contained 24 pieces. If seven pieces were caramels, how much
did the caramels weigh?"

Here is a problem that appears in a fifth-grade arithmetic book currently on the approved textbook list of New
York City:

"How many quart jars will Miss Tomkins need to can
five bushels of tomatoes?"

My feeling is that in the one-to-one relationship between pupil and teacher which is typical of individualization, there is
less likelihood that completely unanswerable problems, such
as the two examples given, will be answered in a rote fashion,
completely without regard to the real world that surrounds
us.

The point I am trying to make is that I do not consider it very important whether or not the individualized
group does better than the conventional group on standardized
tests. In fact, considering the instructional stress on concepts rather than on rote processes, I am always amazed
that the group taught individually does as well as it does on
a test based on another approach.

What I consider more important and what is reflected
in study after study (including many in this book) is that the
individualized group shows a greater understanding of concepts and acquires a much more favorable attitude and feeling

toward mathematics.   This last factor is of prime importance.   Let's face it, mathematics is without a rival in being both feared and hated in the elementary and secondary
schools.   In fact, this dread comes up into college, where
many otherwise good students go to any lengths to avoid required math courses, as long as they can possibly get away
with it.   If, therefore, we have in the individualization of
mathematics instruction a means of diminishing this dislike
or even of turning a negative attitude into a positive one, we
certainly have come upon something sorely needed.

William A. Graham

        The fourth selection in this chapter is from a 1964
Arithmetic Teacher article written by Mr. Graham, who was
at the time of writing this article and is now director of the
Kilby School at Florence State University in Alabama.   Mr.
Graham has long been interested in individualized instruction,
first as a teacher, subsequently as a supervising principal
in the Florence City Schools, and now in his present position.   He writes that much of the instruction in all skills is
done on an individualized basis at Kilby School.

        The brief description of the individualized program in
a fifth grade at Kilby outlines a sound program which had
fine results, even on such tests as I labelled unfair to pupils
who had been taught on an individualized basis.   The readers
will, I hope, note that while the youngsters did well on the
tests that Mr. Graham reports, the teachers felt that the
most worthwhile outcome was the heightened interest in
mathematics on the part of the pupils.

Carolyn C. Potamkin

        It is appropriate that the last selection of this chap-
ter, which also happens to be the last item in our list of
program descriptions before we turn to other considerations
in the next several chapters, should be one that describes
what I consider to be the most common type of individualized
instruction in the American classroom.

        This approach might well be called an "accidental"
one, because it is typically one which begins as a temporary
measure, forced upon the teacher because of a realization of
the need for accommodating marked individual differences

among the children in a classroom, and then ends as a permanent, all-out program of individualization.

Caroline C. Potamkin, who describes this experience during her classroom teaching in a fourth grade in such a charming manner, is now a resource teacher in the Montgomery County Maryland Schools, "working with teachers and children to promote greater individualization of instruction." This article was first published in the Elementary School Journal in 1963. It might well have been written for this month's issue of an educational journal. I think that readers will enjoy the article and profit from a careful reading of it.

In a number of articles that describe individualized programs, mention is made of the fact that such plans require a considerable expenditure of time by the teacher. Anyone who approaches the matter of individualization without a full realization of that fact will soon learn of his error. Actually, however, expenditure of time is a factor in almost any worthwhile teaching program in school or out. I am sure that as the various descriptions have been read and digested, many readers can think of ways that a particular mode of operation could be made to be more efficient and thus consume less time, without in any way diminishing the quality of the program of individualization.

THREE APPROACHES TO THE TEACHING OF MATHEMATICS IN THE ELEMENTARY SCHOOL

Jack R. Fisher

A recognition of individual differences is hardly possible without adjustments in the organization of the school. Attempts to deal with individual differences in the past can be classified into three patterns. They are:

Pattern One: This pattern assumes fixed educational goals in a fixed educational treatment. Individual differences are taken into account chiefly by dropping students along the way. Tests are used to decide which students should go faster and be imbued with aspirations for higher education. The social theory involved is that every child should "go as far as his abilities warrant." The first pattern suggests two

variants: the duration of instruction is offered for an individual by sequential selection and weeding out; and the duration of instruction is altered by training to a fixed criterion.
In both of these variants the educational goal for each student
is essentially the same and the instructional treatments provided for the student are fixed.

   Pattern Two: The adaptation to individual differences
is to determine for each student his prospective future role
and provide for him an appropriate curriculum.   Adapting to
individual differences by this pattern assumes that an educational system has provision for optional educational objectives,
but within each option the instructional treatment is relatively
fixed.

   Pattern Three: The adaptation to individual differences
attempts to teach different students by different instructional
procedures.   Within each of these instructional treatments
there is a minimum fixed sequence of educational goals which
must be mastered.   This pattern of adaptation can be accomplished by providing a fixed instructional sequence with provisions made for removing students from the sequence for
remedial purposes, or providing diagnostic procedures of the
student's competencies--his learning habits, achievement,
and skills--on the basis of which a prescription can be made
for a course of instruction specifically tailored to that student.

The Three Curricula Investigated

   During the past several years there has been several
curricular approaches utilized in the mathematics sequence
in the elementary schools of the Baldwin-Whitehall School
District.   The three approaches which are of major concern
in this study are (1) the standard arithmetic curriculum, (2)
the programmed learning materials approach, and (3) the individually prescribed instruction approach.

1.   The Standard Curriculum

   As part of the on-going process of curriculum development in the Baldwin-Whitehall School District, classroom
teachers, administrators, and an arithmetic coordinator met
in the school year 1962-63 to plan an elementary mathematics
program.   As a basis for content selection, they used Bulletin

233-B, The Elementary Course of Study, as a guide.   This
bulletin grew out of a state-wide participation in curriculum
planning for the elementary schools of Pennsylvania.   It is
a course of study, indicating tenative scope and sequence in
terms of growth levels in the various areas of learning or
divisions of subject matter.

The method of determining the scope of elementary
school mathematics was determined by the content found in
Bulletin 233-B and the textbooks that were adopted by the
school district.   The judgment of what was to be taught was
made by the committee and was to be implemented in the
classroom by means of the arithmetic curriculum bulletin.

In the selection of content, the mathematics curriculum
was divided into major units such as addition, subtraction,
multiplication, etc.   These categories were then subdivided
into sub-aims, making them more specific as to what should
be accomplished in the major divisions.   These subject-matter divisions, in turn, became the units that the teachers
would develop from the textbook and use to teach the pupils.
In developing these units, an attempt was made to establish
a hierarchy of pre-requisite learning which would carry over
from grade to grade, thus establishing a curriculum in which
the arithmetic units would be presented in a spiral approach.
The selection of content thus became concerned with the divisions of mathematics and how these divisions could be implemented through the use of a textbook.   The content became determined by the limitations of school time, the great
variations in the ability to learn, and the range and depth of
available information as found in the adopted textbooks.
Since instruction of the mathematics program was characterized by the use of the textbook, the content of the textbook
determined the objectives of the mathematics program in the
district classroom.

2.   Programmed Learning Materials

During the school year 1961-62, a series of exploratory studies was begun to study the feasibility of developing
a system which would allow the use of programmed materials
to individualize instruction in the standard curriculum.   The
work was begun by using programmed instruction in intact
classrooms.   The intact classroom was one designated as a
classroom or grade level in which the teaching procedures
were oriented around the conventional grade by grade

progression of learning.

The use of programmed instruction offered an instrument that would allow the pupil to proceed with the instruction on the basis of his particular requirements.   The instructional procedure for the pupil would be adjusted according to the rate at which the pupil learns and allow him the opportunity to move ahead, or spend additional time on the subject to be mastered.   A detailed analysis of the work accomplished in the intact classroom is to be found in a research report. [1]

The specific conclusions of each study have been reported in the separate studies.   However, it is felt that the broad implications of these studies should be cited because of their importance in establishing the methodology of the studies conducted in 1963-64:

(1)   There is extensive variation in rate of learning among students when they are given the opportunity to proceed at their own rates with programmed learning materials.
(2)   Pretest scores show that many of the students know the subject being taught and that some students are not ready to learn it.
(3)   Different types of teacher-program combinations in several grades made little difference in student achievement.
(4)   The extent of the correlation between general intelligence and achievement as a result of programmed instruction depends upon the particular program involved.   In general, intelligence appears to be related to the pace with which the student goes through the program.
(5)   Extension of the curriculum with programmed materials, necessarily taking away from time spent in conventional grade-level instruction, produced additional learning without being detrimental to the learning of materials usually taught at that grade level.   In general students required to learn more did learn more.

The second set of studies, which were conducted in 1963-64, was deemed necessary due to the limitation of the relatively inflexible intact classroom situation utilized in the earlier studies.   The course content and the kind of subject matter taught by the programs in the 1963-64 school year were selected on the basis of (a) availability for the particular grade levels involved and (b) subject-matter requirements in terms of student need and student-teacher-community acceptance as determined by the Baldwin-Whitehall school administrators. All of the programs were considered by teachers and administrators as representative of the subject matter normally taught at the grade level in which they were introduced. Of the eight commercial programs used, six concerned mathematics or arithmetic; the other subject matters used were spelling and general science. This study was only concerned with the third-; fourth-; and fifth-grade levels.

The materials were in programmed textbook format and presented the individual student with step-by-step sequences of questions, problems, and answers to which he was actively responding. They also provided feedback or correction to keep the student informed of his progress. Along with continuous active student response, they provided continuous reinforcement, and allowed the student to work at his own rate.

3. Individually Prescribed Instruction (IPI)

In the summer of 1964, the Learning Research and Development Center (LRDC) of the University of Pittsburgh, in conjunction with the Baldwin-Whitehall School District, instituted plans to develop an educational environment characterized by an ungraded curriculum and the individualization of instruction. The instructional program was to be instituted in the Oakleaf Elementary School in September, 1964.

Operational curriculum committees composed of LRDC staff and Oakleaf personnel were organized to study the content areas of mathematics, reading, and science. The primary concerns of the mathematics curriculum were to:

(1) Determine the areas to be found in elementary school mathematics and the scope and sequence involved in each of the areas.

(2) Determine the sequence of behavioral objectives to be taught within an area and between areas.

(3) Break down the particular behavioral objectives into basic units or categories which would represent meaningful teaching units.

(4) Write clearly defined behavioral objectives in such a manner that their accomplishment could be established in an empirical fashion.

(5) Select materials, develop methods, and establish

procedures leading to the accomplishment of the behavioral objectives.

(6)   Select or devise tests to place the students properly on an individually prescribed instructional continuum and to determine the extent to which the objectives have been accomplished.

In developing the IPI curriculum, members of the staff included teachers, subject matter specialists, psychologists, and supervisory personnel.   Various mathematics curriculums were investigated to see what would be included in a curriculum to be used for this program to be consistent with an elementary school curriculum presently in operation.

There was no attempt to select any one series of textbooks, workbooks, or any related materials that would fit any particular curricular approach, either by local school district criteria or any consensus approach.   The procedure used in developing the curriculum was first to define the sequence of objectives for kindergarten through grade six and beyond.   These behavioral objectives became the basis for the selection of the materials to be used in the teaching sequences.

The majority of these materials came from seven commercial sources.   During the first year of the study (1964-65), even though materials were assigned mainly from commercial sources which seemed to fit the IPI continuum objectives, other materials had to be developed to accommodate gaps created by the established objectives.   Throughout the first year of the study, a constant revision and search for better materials was conducted.   This refinement of the curricular materials continued during the summer and became an improved model for the second and third years of the study (1965-66 and 1966-67).

Because of the many problems involved in establishing an on-going innovative program of this magnitude into a school system, it was recommended that the individually prescribed instruction program be conducted in one school, Oakleaf Elementary School.   The comparatively small school population to be dealt with, a single grade level group, and a positively-oriented faculty were all felt desirable in controlling many of the experimental variables in such a curriculum undertaking as this.

(1).   Assumptions Underlying the Standard Curriculum

In the formulation of these assumptions, the author has attempted to determine those elements that have gone into those areas which seem to have manifested themselves in the teaching of elementary school mathematics, relevant to this study.   These appear to have served as the guidelines, by the teachers, in the implementation of the mathematics curriculum in the classes defined as the standard arithmetic curriculum classes.

(1)   In the accomplishment of objectives the teacher must understand the developmental stages of child growth which will enable her to meet the needs and capacities of the individuals and to bring the pupil into contact with the problems and challenges which will facilitate the exploration of his interests.

(2)   In the instructional process, the teacher has a command of the subject matter she is teaching and is able to select those instructional materials, learning experiences, and arrange the classroom conditions which will satisfy the child's needs and capacities for his development.

(3)   The teacher can make extensive provisions for individual differences in the standard classroom by subgrouping within the class, a differentiation of pupil assignments as determined by their abilities, and the employment of materials relevant to the capabilities of the pupils.

(4)   The teacher must determine within the range of the existing experiences of the children those areas that have the potential of presenting new experiences that are conducive to further growth.

(5)   Although there are individual differences within the standard classroom, the pupils are basically enough alike in factors such as chronological age, needs, and interests, and a similar mental frame of reference that an instructional setting can be presented that would benefit all of the pupils.

(6)   Through the selection and organization of subject matter, the learning experiences can be controlled and directed by the learner and guided by the teacher.   This emphasizes the building of habits and skills as integral parts of extending growth.

(7)   A course of study is to be regarded as a guidepost, and available materials something yet to be adapted and adjusted to the individuals, perhaps not tailor-made, but

susceptible to alterations to meet the requirements of the individual.

(8)   The educative process requires the child to experience the working out of the problem situations leading to the acquisition of skills and abilities out of which learnings come.   The problem situations or the goals to be sought must be within the capacity of the individual to reach.

(9)   It is important for the teacher to provide a classroom atmosphere that will encourage all students to become active participants in a large group instructional program. This would include profitable learning experiences for that segment of any class that could learn through listening, even though not actively involved.

(10)   The elementary school is conceived as a non-specialized institution, in that it provides the same type of training to all throughout the period of attendance.

(11)   The curriculum is planned in advance, but with much opportunity for pupil participation in self assignments, discussion of findings, and evaluation of their own work.

(12)   The scope of work must be defined, the time planned, materials made available, and classroom needs considered to prevent good learning from being hindered by poor management.

(13)   A generous supply of varied and up-to-date instructional materials and supplies are available to the children, enabling them to accomplish a learning task.   Textbooks, both basic and supplementary, on several levels of difficulty are provided for the children, because of the varying individual differences found in any classroom.

In studying the development of the curriculum guide, it becomes apparent that the objectives and sub-aims of the guide had to conform in a rather rigid sense with the material found in the textbooks to be used.   This would agree with Cronbach's conclusion that "at the center of the present-day educational scene in America is the textbook.   It takes a dominant place in the typical school from the first grade to the college. "[2]   The textbook used, compatible with the objectives of the curriculum guide, was Growth in Arithmetic by Clark, Junge, and Moser.   This particular text was in use up to and through the school year 1962-63.   At this time it was felt that this series was not emphasizing enough of the "modern" concepts felt desirable in the teaching of mathematics.   The textbook finally selected for this task was the series Modern Arithmetic Through Discovery by Morton, et al.

A decision was reached by the mathematics curriculum guide committee to retain the present classroom guide and continue to implement it through the newly adopted textbook series.   In this series the authors have drawn from several exploratory programs in elementary mathematics and have selected from them ideas which were incorporated into this textbook series.   Instead of calling their content subdivisions chapters or units, they have chosen the term "learning stages. "

As mentioned previously, the standard mathematics program was organized around the guide, "Arithmetic Curriculum, " and the adoption of a single textbook series.   The authors had as a goal the establishment of new standards for the teaching of arithmetic, at the same time providing pupils with a series of challenging, meaningful, and appropriate learning experiences.   The authors carefully examined several exploratory programs in elementary mathematics.   These included programs of the following: the School Mathematics Group; the Syracuse University "Madison Project"; the University of Illinois Committee on School Mathematics; the University of Maryland Mathematics Project; and the Greater Cleveland Mathematics Program.   From all of the topics that might be taught, the authors selected those topics which, in their judgment, could be taught most appropriately at each grade level.

The program was developed to emphasize inquiry, exploration, and discovery.   The texts were designed to help pupils move from one idea to another; and throughout this procedure, the pupils are encouraged to explore and discover relationships.   There were teacher editions provided, with answers and marginal teaching suggestions.   New content and approaches are explained in these marginal notes.

(2).   Assumptions Underlying Programmed Learning

In the use of programmed learning materials in the classroom, it has become necessary to explore and determine exactly what the assumptions would be in developing a program in elementary school mathematics involving changes in the utilization of such materials, differentiation of the teacher's role, and the procedures by which the pupils would become involved in the learning experiences.   These assumptions characterize those elements of programmed learning that were felt to be important for the accomplishment of adequately dealing with the individualization of the instruction.

(1)   In the use of auto-instructional aids it is possible to relieve the teacher of many of the clerical tasks in which she is involved, thus freeing her to engage in the instructional tasks which are relevant to children's progress.

(2)   A teacher can supervise an entire class using programmed materials at one time and still permit each child to work and progress at his own rate.   Attention can be given to the problems of individual children as the need arises.

(3)   A hierarchical arrangement of the subject matter must be accomplished, so that the learner will be able to master the simpler skills before proceeding to those that will become more complex.

(4)   The programmed learning materials are structured into steps small enough for the learner to learn rapidly and retain that which has been presented.

(5)   It is of importance to determine the student's level of understanding, or entering behavior, of the subject matter before beginning his educational experience.

(6)   With the individual interacting with programmed learning materials, the schedule of reinforcing opportunities is built within the confines of the program which has been designed for him.   Relying on the differential reinforcements emanating from the teacher is based on large amounts of verbal behavior and may be faulty.

(7)   A concern in programmed learning instruction is that of providing an immediate confirmation or knowledge of results about the correctness of the learner's responses, since no teacher need be present to mediate between the learner and the teaching device.

(8)   Since it is impossible for a teacher to deal capably with the needs of all the children at one time, programmed materials do much to increase the teacher's control over the learning activities in a classroom.   By presenting educational content and permitting the teacher to be the mediator of the learning process, some of the advantages of tutorial instruction are inherent in programmed instruction.

(9)   In the preparation of materials for programmed instruction, the programmer, or author, is required to think in terms of student responses.   This forces him to think in terms of the exact behavior he is trying to teach children, how it relates to pre-requisite skills, and how it will

determine future learning.

(10) Ineffectiveness of classroom procedures is often attributed to "short attention span" or lack of concentration. The principles of programmed learning require the learner to be an active participant. If temporary lapses of action occur, the teaching material does not pass by, but is reactivated when the learner returns as a participant.

(11) Programmed learning materials should be recognized as being tutorial in nature, inasmuch as they present the learner with a series of problems which require an active response on his part.

(12) In utilizing programmed materials in the classroom, the teacher would be able to plan more effectively her instructional role in dealing with the students by:

(a) Knowing exactly where all children were in the instructional sequence.

(b) Examining problem areas that would need additional clarification for the individual.

(c) Eliminating repetitious drill and review.

(d) Devoting class time to further development of concepts.

(e) Possibly revising the program, based on empirical evidence.

(f) Allowing for varying rates of learning and achievement.

(g) Proceeding to a higher level after manifestations of mastery of essential elements of learning.

(13) The self-instructional devices utilized in programmed learning are designed to take the learner from a low to a high level of proficiency in a subject matter content area.

(14) It is especially desirable in utilizing programmed learning materials as instructional devices that the learner will not only know how to respond correctly to the content of the program but will be able to deal effectively with related materials and concepts, thus enabling a transfer from a training situation to one of application in a task situation.

(15) Programmed learning instruction is an attempt to make education more efficient because of the demands of an ever-growing population and the needs of a more complex society. It provides material to the student in order to make him a more successful and efficient learner.

The research program for the school year 1963-64 was developed around the task of providing as much individualization of instruction as possible, with the aim of providing for the acceleration and extension of instruction. The studies reported herein involve the third-, fourth-, and fifth-grade levels.   In an attempt to adapt the pace of instruction to the individual student, programmed instructional materials were utilized at various grade levels within a classroom structure which was revised in order to permit a flexible teaching situation.

As part of the program development, the teachers to be involved in the various studies were invited to work with staff members of the Programmed Learning Laboratory of the University of Pittsburgh during the summer preceding the opening of school.   The major effort during these training sessions was to anticipate any problems that would be encountered, the writing of manuals, and the procedures and strategies necessary for a successful operation of an experimental study.

A great deal of educational research has been undertaken in attempting to solve problems encountered in education.   One approach to the study of this problem is to begin by studying the components of the system to be changed and to consider the difficulties of producing changes in these components.   Travers[3] related that the two major components are behavior systems and consist of the behavior systems of the student and the behavior systems of the teacher.   Educational research based on the assumption that teachers will change their behavior if they are told how to make their teaching more efficient is not likely to occur.   It is the component which is least amenable to change.   It is the behavior of the pupil which the whole system exists to modify. In this series of studies an important process occurs in the educational system: the teacher component of the system delegates control to the equipment components of the system. Equipment, in this situation, includes such items as textbooks, workbooks, programmed textbooks, and the materials involved in the individually prescribed instruction curriculum.

In this discussion on the components of an educational system, the area considered most amenable to change was in the realm of instructional materials and devices which do not require major changes in the habit structure of teachers. Since most teachers find it feasible to use instructional materials in the classroom, it was felt that some measure of success in the educational process could be accomplished by introducing programmed materials as a means of individualization.

The major purpose of the series of studies conducted
with the intact classrooms was to examine student instruction
and achievement in basic subjects taught largely by programmed instruction in the elementary school and to determine what possibilities might be suggested for the improvement of instructional practices in the classroom.

As a result of the work done with the intact classes,
it became apparent that the individualization features of programmed instruction could not be realized unless the intact
classroom changed its organization to permit a more flexible
progression. These results suggested that an important need
in the improvement of classroom instruction is the development of procedures which more precisely determine what a
pupil knows when he comes into the instructional situation,
and a procedure for adapting the instructional sequence to
his particular requirements. It was felt that by using programmed materials, restructuring the classroom procedures,
and developing individualization techniques it would be possible to increase individualization of instruction and provide
opportunities to extend learning.

Programmed instruction was utilized as one possible
answer to enhancing the advantage of the public school mass-
education, while at the same time reinstating some of the
advantages of individual student-teacher interaction. It is
primarily concerned with the precise selection and arrangement of educational content based upon what we know about
human learning. One of the major problems posed is that
of the supposed impact on the dehumanization of learning.
A program for a teaching machine can be just as personal to
the learner as any other instrument used to share verbal information. Bruner[4] suggests that these devices exist to aid
the teacher in extending the student's range of experience,
in helping him understand the underlying structure of the
material he is learning, and in dramatizing the significance
of what he is learning. These devices can also lift some of
the load of teaching from the teacher's shoulders. The most
interesting part of the problem is to find out how those aids
can accomplish this task.

In presenting the term "instruction" in a programmed
learning sense, it becomes necessary to qualify it. Instruction would refer to any specifiable means of controlling or
manipulating a sequence of events to produce modifications of

behavior through learning.   Lumsdaine[5] determines instruction as being applicable whenever the outcomes of learning
can be specified in sufficiently explicit terms to permit their
measurement.   The definition could be carried a step further
to make it one of auto-instruction.   Here, according to
Stoops and Marks[6], auto-instruction is a process involving
the use of carefully planned materials designed to produce
learning without necessarily requiring the immediate pres-
ence of anyone other than the pupil.

     The usual textbook does not control the behavior of
the learner in a way which makes it highly predictable as a
vehicle of instruction or amenable to experimental research.
The fact that it does not in itself generate a describable and
predictable process of learner behavior may be the reason
why there has been very little experimental research on the
textbook.

     Additionally supporting the reason why programmed
instruction was selected as the instrument for implementing
the task of individualization are the statements of Lee and
Lee.[7] It is their thought that all learning should be conceived in large units, the meaning of which each child should
be aware.   The large unit should be divisible into smaller
units, still meaningful, and having an understood connection
with the larger unit.   It should be made certain that all of
these units, or difficult portions, are mastered and combined
into the meaningful whole.

     As has been indicated by Skinner, a programmed de-
vice acts like a private tutor in several respects:

     There is a constant interaction between the program
and the student; the student is always alert and busy;
     like a good tutor, the program requires that a given
point be thoroughly understood before the student moves on;
     the program presents just that material for which the
student is ready;
     the program helps the student to come up with the
right answer; and
     like a skillful tutor, the program encourages and rein-
forces the student for every correct response. [8]

     In addition to its use as a teaching aid, programmed
learning can be used to investigate laboratory findings of the
science of learning to educational psychology.   One of the
goals of this study is to attempt to develop teaching techniques

208   Individualized Instruction in Mathematics

based on a scientific foundation of learning theory.

As previously indicated, the processes of teaching and learning can be made a matter of scientific study, on the basis of which a more efficient model of instruction can be developed. It is clear that with the identification of instructional variables, which influence the modification of behavior through instruction, it will be possible to improve education effectively. Through this study, it becomes increasingly evident that instruction and learning are amenable to systematic description and improvement through experimental inquiry.

As the study proceeded, it became apparent that many of the students were completing the programmed materials ahead of an anticipated schedule, and that there was a need for advanced work in particular topics. As a result, it was necessary to develop supplementary self-teaching materials so that the individualization procedures could be continued.

(3). Assumptions Underlying Individually Prescribed Instruction (IPI)

In the developmental phases of the Individually Prescribed Instruction curriculum, it became necessary to provide a number of assumptions that would provide adequate guide lines for all of those involved with the project. These were especially crucial for the establishment of behavioral objectives, learning materials in the instructional sequence, procedures to be implemented in the classroom, and the testing and evaluation of the pupil's progress. The following statements by Lindvall and Bolvin became the specific assumptions of the project.

One obvious way in which pupils differ is in the amount of time and practice that it takes to master given instructional objectives.

One important aspect of providing for individual differences is to arrange conditions so that each student can work through the sequence of instructional units at his own pace and with the amount of practice he needs.

If a school has the proper types of study materials, elementary school pupils, working in a tutorial environment which emphasizes self-learning, can learn with a minimum amount of direct teacher instruction.

In working through a sequence of instructional units,

no pupil should be permitted to start work on a new unit until he has acquired a specified minimum degree of mastery
of the material in the units identified as pre-requisite to it.

If pupils are to be permitted and encouraged to proceed at individual rates, it is important for both the individual pupil and the teacher that the program provide for frequent evaluations of pupil progress which can provide a basis
for the development of individual instructional prescriptions.
Professionally trained teachers are employing themselves most productively when they are performing such
tasks as instructing individual pupils or small groups, diagnosing pupil needs, carrying out such clerical duties as
keeping records, scoring tests, etc.   The efficiency and
economy of a school program can be increased by employing
clerical help to relieve teachers of many non-teaching duties.
Each pupil can assume more responsibility for planning and carrying out his own program of study than is permitted in most classrooms.
Learning can be enhanced, both for the tutor and the
one being tutored, if pupils are permitted to help one another
in certain ways. [9]
From these assumptions, Glaser has provided us with
a definition of this innovation in instructional practice: "Individually Prescribed Instruction is a unique plan for arranging and carrying out classroom instruction.   Its goal is
to provide an educational environment which is highly re-
sponsive to differences among children. " As the project
continues into the third year, additional assumptions have
been enumerated by Dr. John O. Bolvin, the IPI Director.
They are:

(1)   A sequence or ordering of objectives can be made
in each of the curriculum areas.
(2)   Learning to be meaningful must be placed to
some degree in the hands of the learner.
(3)   Active responses are better than passive responses
for most learning.
(4)   Student errors when used for diagnostic purposes
are not punishing.
(5)   In the elementary years of school, all children
can move through the same objectives in the tool subjects.
(6)   Entering behaviors can be measured and suitable
instructional materials can be prescribed for each child.
(7)   Permitting the child to work at the boundary between what he knows and what he needs to know next is itself
a motivating strategy.

(8) If individualization is to be accomplished then children must be provided the materials that permit self-learning.

(9) Children can learn effectively with much less teacher verbal instruction.

(10) Children can learn from other children (peer-tutoring).

(11) Not all teachers can work effectively with all students.

The goal of the Learning Research and Development Center is to create a model educational environment with an ungraded curriculum which can be tailored to the instructional needs of the individual student. With the present knowledge we possess concerning individual differences, we can proceed to apply this knowledge to the systematic management of individual learning in an appropriately designed school system. An individualized system needs to assess each student's progress, so that appropriate decisions can be made in keeping with his achievement, capabilities, and interests.

It seems appropriate at this time to cite the major objectives of the project, not in the subject matter context of arithmetic, but in the over-all goals to be implemented through the subject matter areas. It should be evident that these should not be considered a dichotomy, but an indivisible whole directed towards the individualization process.

To restate certain elementary school curricula in terms of a continuum of specific behavioral objectives, for the purpose of monitoring and assessing the progressive development of each child's competence in subject matter areas.

To provide a variety of materials and techniques of instruction to meet the individual needs of students.

To establish teacher functions and procedures to facilitate individually prescribed instruction.

To develop a school structure and organization that permits the flexibility required for individualized learning.

To work toward operating procedures which are within the financial means of most innovative schools.

The major assumption to keep in mind is that the individually prescribed instructional curriculum (IPI) should basically be designed to produce an effective and workable plan for the utilization of certain procedures in providing for individual differences among students. The persons involved

in and responsible for this progress were encouraged to develop these desired changes for achieving the goal of individualization.

Results

In the investigation of the three instructional treatments, the individually prescribed instruction and the programmed learning instruction have made some innovative departures from those of the standard classroom instruction
curriculum.   Those pupils involved in these two innovative
programs appear to have done as well as the standard classroom instruction pupils, as evidenced on the sub-tests of the
various forms of the Metropolitan Achievement Tests.

This, in spite of the fact that the nationally normed
achievement tests are geared to analyzing the achievement in
school systems consisting largely of graded structure.   This
would naturally require the tests to be written within a
framework of what is typically taught in a particular grade.
This type of testing would, thus, not lend itself well to a
curriculum designed for the individualization of instruction.

In the programmed learning instruction curriculum,
it was unfortunate that there were not sufficient programs
written to cover the full scope and sequence of what is
typically found in the arithmetic content of the usual elementary school program.   This, again, is a serious obstacle
in developing a comprehensive elementary school arithmetic
curriculum.   Even though these materials lend themselves to
some individualization, this lack of scope handicaps comparing them to the standard classroom instructional curriculum.

The IPI curriculum has as its base the mastery of
behavioral objectives, arranged in a hierarchy, rather than
the "covering" of a pre-determined graded content area.
This may also serve as somewhat of a handicap in making
comparisons with a standard classroom curriculum, as the
pupils may have to spend longer amounts of time and participate in modes of instruction other than the simple use of
a textbook in order to reach what is considered a mastery
of the behavioral objectives.   The major emphasis in the
IPI curriculum is to diagnose the deficiency of the pupil in
the behavioral objective, then prescribe an instructional sequence that will enable him to manifest mastery of the objective.

This is quite different from teaching a group of children specific arithmetic content that leads to satisfactory scores on an achievement test peculiar to what he has been taught.   In the IPI curriculum, the program has not been specifically oriented towards preparing pupils to score higher on the standardized achievement tests, as compared with curriculum treatments which appear to use these tests as evidence of their success.   The IPI curriculum, as it continues to develop, should offer evidence of pupils' ability to make higher achievement results on the standardized tests and still retain what appears to be of greater importance, the ability to exhibit greater proficiency in arithmetic skills at a higher mastery level.

The data found on the Iowa Basic Skills Multi-Range Test Tables offers considerable evidence that there is a need for an instructional method by which the pupils at any grade level can extend themselves into doing work beyond that in which they are confined by grade placement.   The procedures, instructional materials, and diagnosis of pupil difficulties through testing, found in the individually prescribed instruction curriculum, offer a compatible solution to this problem.   In the implementation of the IPI assumptions, a greater impact can be made to develop the potential of the individual.

Summary

Assumptions

(1)   In the assumptions of the three curriculum treatments, there exists some commonality of purpose relevant to the mathematics achievement of elementary school pupils. However, there are certain assumptions peculiar to the individual curriculum treatments that are more likely to accomplish the goals of individualizing instruction.

(2)   By means of such techniques as grouping and a differentiation of assignments and materials, some individual differences can be dealt with in any of the curriculum treatments described.

(3)   A course of study, a programmed sequence of materials, and a continuum of mathematics objectives were utilized in providing continuity in the mathematics areas and were deemed necessary in providing guide lines for the scope and sequence of the mathematics programs.

(4)   The teacher was considered essential in controlling and directing the active participation of the pupil in all of the curriculum treatments.

(5)   The individually prescribed instruction curriculum contained assumptions which considered the importance of the teacher as a diagnostician of the specific learning difficulties of the pupil.

(6)   The assumption which states that an individual instructional prescription will be written for each pupil, after an evaluation by the teacher of his entry behavior into a learning situation, as determined by a test score, and allowing him to proceed at his own rate with a mode of instruction suitable to his needs is to be considered as a unique contribution of the individually prescribed instruction curriculum.

Objectives

(1)   The objectives are not equally clear in stating specifically what the pupil should manifest after an instructional sequence.   With the exception of the individually prescribed curriculum, they are not stated in behavioral terms and thus are not able to be substantiated on the basis of empirical evidence.

(2)   An arrangement of objectives is to be found in all three curriculum treatments, but only in the individually prescribed instruction and programmed learning curriculums were certain pupil-performance criterion scores required to insure objective mastery before the pupil went on to a more difficult objective.

(3)   The individually prescribed instruction curriculum contained a hierarchical arrangement of behavioral objectives requiring a close integration of materials and instructional procedures necessary to accomplish the objectives that were not present in the other two treatments.

General Methods

The general methods employed by the teacher reflect the orderly or systematic way in which the teacher acts as the mediator of instruction in accordance with a definite plan.   The three different curriculum treatments required differences in the role and function of the teacher as dictated by the distinctive elements of the program.

The methods used in the individually prescribed

instruction and programmed learning programs were determined by pupil product criteria rather than teacher presentation of mathematics content.

Materials

The characteristics of the instructional materials
largely determine how they will be used in the classroom.
Their utilization implements the assumptions and objectives
of the curriculum. The standard classroom employs a textbook geared largely to group teaching practices; the programmed learning materials provide for some individualization
of rates of learning, but do not provide adequate paths, techniques, and materials; the individually prescribed instruction
materials have been designed to have pupils achieve mastery
of concepts found in the behavioral objectives of an all-inclusive elementary school mathematics continuum.

Classroom Procedures

(1) The three curriculum treatments required different kinds of lesson plans to be prepared by the teacher.
The standard classroom emphasized specific textbook content, thus requiring the teacher to develop procedures to
teach for content coverage in a group situation.   In developing procedures for the programmed learning curriculum, the
teacher became involved with the status of the individual pupil within the program.   The teacher's function becomes
more tutorial in nature.   In the individually prescribed instruction curriculum, the tutorial emphasis becomes greater
as the teacher now becomes involved with the diagnosis of
learning difficulties, the pupil acquisition of behavioral objectives, the analyzing of empirical data, and the writing of
instructional prescriptions.

(2) The planning of daily lessons and conducting
classroom procedures is largely dependent on how well the
teacher can monitor the involvement of the pupil with his
instructional tasks.   The instruments available to the teacher
in the individually prescribed instruction curriculum provide
a greater availability of feedback of information to the teacher, concerning the pupils' progress, than would be found in
the other two treatments without a considerable amount of
time and modification of classroom procedure taking place.

(3) The lesson plans and procedures of the individually prescribed instruction and programmed learning curriculums are determined by the use the teacher makes of the
testing and evaluation devices.   These devices become relevant to the instruction of the pupil.   In the standard classroom they are used mainly as ex post facto measurements of
pupil achievement and become translated into a report card
grade.

Standardized Testing and Data

The programmed learning and individually prescribed
instruction curriculum pupils have been involved with two instructional approaches that differ in many respects from that
of the standard classroom.   Those pupils participating in the
innovative programs appear to have done as well as the
standard classroom pupils, as evidenced by the various
graded forms of the Metropolitan Achievement Tests.   Since
these tests have been designed for school systems having
graded structures, this did provide evidence that pupils involved in programs of individualized instruction can do as
well on tests not specifically suited to programs with non-
graded characteristics.

The data collected from the Multi-Range Iowa Basic
Skills Tests strongly suggests that, in all three curriculum
treatments, there is a need for an instructional method by
which pupils at any grade level can extend themselves into
doing work above and below the grade at which they have
been placed.

Notes

1.   Robert Glaser et al.   Studies of the Use of Programmed
     Instruction in the Classroom.   Technical Report 1.
     Learning Research and Development Center, Univer-
     sity of Pittsburgh, 1966 (mimeographed).
2.   Lee J. Cronbach.   Text Materials in Modern Education.
     Urbana: University of Illinois, 1955; p. 216.
3.   Robert M. W.   Travers.   "Training Research and Edu-
     cation. " A Study of the Relationships of Psychological
     Research to Educational Practice, edited by Robert
     Glaser.   Pittsburgh: University of Pittsburgh Press,
     1962; Chapter 17.
4.   Jerome S. Bruner.   The Process of Education.   New

York: Vantage Books, 1960; p. 84.
5.   A. A. Lumsdaine.   "Instruments and Media of Instruc-
     tion. " Handbook of Research on Teaching, edited by
     N. L. Gage.   Chicago: Rand McNally, 1963; p. 586.
6.   Emory Stoops and James R. Marks.   Elementary School
     Supervision: Practices and Trends.   Boston: Allyn
     and Bacon, 1965; p. 261.
7.   J. Murray Lee and Doris K. Lee.   The Child and His
     Curriculum.   New York: Appleton-Century-Crofts,
     1940; p. 186.
8.   B. F. Skinner.   "The Science of Learning and the Art
     of Teaching. " Harvard Educational Review 24(2):29,
     1954.
9.   C. M. Lindvall and John O. Bolvin.   The Project for
     Individually Prescribed Instruction.   Oakleaf Project.
     Working Paper 8, Learning Research and Development
     Center.   University of Pittsburgh, February 1966.

NEW MATHEMATICS CURRICULUM IS PART OF A "QUIET REVOLUTION" IN TEACHING METHODS

[from a publication of the]
Regional Education Laboratory for the Carolinas and Virginia

✔ Children learn more easily when the skills they
have to master are presented as a logical sequence of small
tasks. ✔ Children find that learning can be fun when they
are offered a variety of ways to learn, and when they can
exercise some direction over how they learn. ✔ Teachers
are most effective when they can work with pupils on an individual basis, or with small groups of students.

These are some of the basic assumptions about learning which are incorporated into an innovative mathematics
curriculum developed by the elementary school division of the
Regional Education Laboratory for the Carolinas and Virginia
(RELCV).

Dr. Frank Emmerling, director of RELCV's elementary school division, describes the new curriculum--called
Individualized Mathematics System (IMS)--as "part of the
quiet revolution that is drastically changing the way students
learn--the individualized instruction movement. " IMS is not

a textbook or a workbook, but a series of about 8,000 "teaching pages." Each page is laminated in pastic, and can be erased and used over and over again.

In February 1970, eight schools began using the curriculum on an experimental basis. Nearly 1,000 second- and third-grade pupils are involved in the year-long field-testing.

IMS divides the elementary school mathematics curriculum into 11 topics: numeration, addition, subtraction, multiplication, division, fractions, mixed operations, money, time, systems of measurement, and geometry. Each of the 11 topics is divided into ten levels of difficulty. In each of these ten levels, there are a number of skills that must be learned. For each skill, there is a folder of from four to 12 teaching pages. "By dividing the work into skill units," Emmerling said, "mathematics is presented as a sequence of small, related tasks which the average elementary-school pupil can master on his own in the classroom."

As they begin using the IMS curriculum, students take a placement test. From the results of this test, the teacher develops a "student profile chart." This profile tells each student what level of difficulty he should begin work on for each of the 11 topics.

The student begins work on the skill folder the teacher prescribes for him, using a special pen to ink in his answers. When he completes the folder he takes a "check-up" test, which is the last page of every folder. If he passes, he records his score, erases his pages, and gets the next folder. If he fails, the teacher will work with him individually or give him a special assignment dealing with the problem areas.

IMS was developed under the mandate of the regional education laboratories to "develop and demonstrate a rich array of tested alternatives to current educational practice." Like most products of the 15 laboratories across the nation, IMS was designed to meet regional needs--yet might be used effectively in schools across the nation.

INDIVIDUALIZATION IN GRADED
AND NON-GRADED CLASSROOMS

Bernard R. Wolff

It was the purpose of this study to determine if significant differences do exist in the individualization of arithmetic instruction between graded and non-graded classrooms.
A packet of Q-sort cards was sent to each non-graded and
graded teacher in the sample. The Q-sort consisted of 70
cards, each with one statement describing an elementary
classroom arithmetic instructional practice.   These statements are given below; the sequential order is for the convenience of the reader.   When mailed to the teacher, the
cards were in no special order.   The statements were
printed on "3 X 5" cards.

(1)   Arithmetic groups are formed daily on the basis
of common student difficulties.
(2)   The class is divided into several groups for
arithmetic instruction.
(3)   Students who complete their assignments before
the end of the arithmetic period are given work options in
other classroom subjects.
(4)   Some arithmetic topics may be omitted for some
students if the need arises.
(5)   Diagnostic arithmetic tests are used primarily
with the slow student.

(6)   Students are permitted to work on arithmetic assignments at times in addition to the arithmetic period.
(7)   On the basis of their arithmetic ability students
are assigned to other teachers during the arithmetic period.
(8)   A very structured arithmetic scope and sequence
is provided for some students.
(9)   The teacher is provided with several current
arithmetic series from which to plan and use.
(10)   Initially, students are assigned to the classroom
on the basis of arithmetic ability.

(11)   The major criterion for determining group or individual arithmetic placement for instruction is the IQ test
score.
(12)   Students are provided supplementary (non-text)

arithmetic materials of varying interest and ability levels.

(13) Self-instructional materials are available for student use.

(14) A definite period each day is designated for arithmetic instruction.

(15) The teacher presents the same arithmetic topic to the whole class but varies assignments in terms of depth and length.

(16) Students are given opportunities to work in basal arithmetic texts specified for other grade levels.

(17) Students are required to do a minimum number of problems on each topic presented in class.

(18) The teacher uses diagnostic tests as the need arises.

(19) Different "seat-work" assignments are given to groups and individuals.

(20) Diagnostic tests are administered to the whole class at the beginning of the school year.

(21) Group or individual arithmetic instruction is based on an informal arithmetic inventory.

(22) Students use a variety of manipulative materials in learning arithmetic concepts and skills.

(23) In addition to regular instruction groups, some students are grouped occasionally on the basis of various arithmetic interests.

(24) Arithmetic topics are presented to individual students or small groups as they are ready.

(25) Class size has been considered in planning for the arithmetic instruction.

(26) Approximately once a week, a teacher conference is planned with each student.

(27) A remedial arithmetic class is provided for students with extreme arithmetic learning problems.

(28) Each student selects his own instructional materials in learning various arithmetic concepts and skills.

(29) Arithmetic instruction is integrated with other subjects, such as social studies and science.

(30) A resource teacher or special personnel work some of the time with the more able arithmetic students.

(31) Students participate in setting up standards for evaluating the quality of their own work.

(32) Each student turns in his written arithmetic assignment each day.

(33)   Arithmetic achievement test scores are the major criteria for the placement of students in groups within the classroom.

(34)   A percentage of number right score is recorded for each written assignment completed by the students.

(35)   All arithmetic assignments are corrected by the teacher.

(36)   Students spend part of the arithmetic period working at the chalkboard.

(37)   Students are allotted approximately the same amount of time each day for arithmetic instruction.

(38)   Students who do not finish their arithmetic assignment during the regular arithmetic period are encouraged to use "free class time" or may take it home.

(39)   The same "seatwork" assignments are given to the whole class.

(40)   Students work on the same page-by-page assignments at the same time in the same basal arithmetic text.

(41)   Students use the same supplementary (non-text) arithmetic materials.

(42)   Review and drill activities are conducted with the whole class.

(43)   Weekly arithmetic tests are given to the whole class but are differentiated in terms of length and difficulty.

(44)   Students who complete their assignments before the end of the arithmetic period are given additional work in the same topic.

(45)   Enrichment activities are provided only for the more capable students.

(46)   Students who complete their assignments before the end of the arithmetic period are given supplementary (enrichment) material from which to choose.

(47)   All students participate in class arithmetic games and competitive contests.

(48)   The teacher presents each arithmetic concept, topic, or skill in a variety of ways.

(49)   Rooms are provided of sufficient size to allow flexible arrangements of seating and work areas.

(50)   Students who complete their assignments before the end of the arithmetic period are given the assignment for the next day.

(51)   The district guide for arithmetic is the primary source for the scope and sequence experiences for the

students in the classroom.

(52)   Checksheets for recording the progress in arithmetic skill and concept development is kept by the teacher
for each child.

(53)   Each student keeps a record of his progress in
arithmetic development.

(54)   Students work on arithmetic assignments growing
out of their classroom and other real-life experiences.

(55)   Make-up groups and conferences are provided
for those students who have fallen behind.

(56)   Initially, students are assigned to the classroom
on the basis of reading ability.

(57)   Answer sheets are provided for independent student correcting of assigned work.

(58)   Self-administering and correcting tests are provided for the students.

(59)   Classes or groups for instruction in arithmetic
are based primarily on the social maturity of the students.

(60)   Teachers use the team approach during the
arithmetic period.

(61)   Students are permitted to help one another during arithmetic periods.

(62)   Review and drill activities are provided for each
group or individual as they complete topic assignments.

(63)   Students work on different assignments in the
same basal arithmetic text.

(64)   Students work on different assignments in different texts but at the same grade level.

(65)   A grade equivalent of 70 per cent is required on
all written work and tests.

(66)   The "Reporting to Parents" procedures require
specific statements relating to the arithmetic skills and concept development of each student.

(67)   Each arithmetic concept, topic, or skill is presented in the manner prescribed by the teacher's manual for
the adopted series.

(68)   Achievement tests are given at the beginning of
the school year.

(69)   Teacher-made mastery tests are given to the
whole class approximately every six weeks.

(70)   The same weekly test is given to the whole
class.

In order to compare the degree of agreement between

the non-graded and graded groups of teachers, each participant was asked to sort the cards in such a way as to differentiate statements which characterized individualization of
arithmetic instruction from "most like" to "least like. " By
asking each teacher to sort the statements, in this case,
ten cards in seven piles, the teacher must use her own
frame of reference, her own conceptual basis for choosing.

This is superior to usual rating scales since it asks the
evaluator to judge items in a relative sense (that is, the
relevance of one item compared with others), while the
rating scales demand that items be judged in absolute
terms; the latter can well lead to spuriously low agreement since different raters, whose evaluations are apt to
be correlated with each other may adopt different anchor-
ing points for their ratings. [1]

CARD SORTING DIRECTIONS

The following statements on the enclosed cards describe practices and procedures in arithmetic instruction that
might be found in an elementary classroom having eight- and
nine-year-old students.   Your task is to sort the statements
into seven piles using ten cards per pile.   Your decision
should be based primarily on what you think would be best
whether or not it is practical for you at this time in your
present classroom situation.   The statements that best characterize "Individualization of Arithmetic Instruction" are
placed in pile 1.   The remaining statements are to be placed
in the remaining piles in terms of next best to least best
keeping individualization the major criterion, until there are
ten cards in each of seven piles.   (The numerals on the
cards have no significance.)

Suggested Procedures

1.   Select ten cards for each pile, 1, 2, and 3 in
that order; then start with pile 7, then 6, and then 5, placing the remaining cards in pile 4.
2.   To facilitate the mechanics of the process and for
easy reviewing of the cards during the selecting; you might
use a desk top or suitable flat surface; place the pile number cards along the top; then place the statement cards below
the pile number cards in a type of solitaire arrangement.
3.   When finished, place a rubber band around each

set of ten cards with the pile number on top of each pile and mail in the enclosed envelope.

Did you find any ambiguous or vague cards? Which ones?___
What was your approximate sorting time?_____

The frequency and card distribution for each teacher sort, based on her interpretation of individualization on arithmetic instruction, would appear as follows:

	Most Characteristic						Least
Pile No.	1	2	3	4	5	6	7
Frequency	10	10	10	10	10	10	10

This rectangular distribution of cards forces more discriminations at the ends of the continuum than a sort requiring a more normal distribution of cards and this uniform distribution "... is an easy requirement ... to force individuals to make discriminations they would not otherwise make. "[2]

A trial sorting was made by several primary teachers not in the research groups. Their reactions and suggestions were used to modify directions and clarify several statements. Though Q-sorting is not a common research tool, Kerlinger believes it to be a valuable instrument particularly when used with "... small sets of individuals carefully chosen for their 'known' or presumed possession of some significant characteristic or characteristics. "[3]

During the months of March and April 1967, visits were scheduled with each of the 20 teachers to observe each classroom situation and to strengthen the reliability of teacher responses during the interview. From 30 to 40 minutes were spent during the arithmetic period observing both the teacher and students. Preceding each observation, a 20- to 30-minute interview was conducted to investigate the extent of individualization of arithmetic instruction practiced by each teacher.

To obtain the general arithmetical variability of the students with each classroom, the Contemporary Mathematics Test was provided each teacher along with manual for administering the test. The tests were administered by the teachers but not scored by them.

Q-Sort Analysis

A Q-sort procedure was used to require teachers in both graded and non-graded classrooms to examine 70 statements printed on "3 x 5" cards. These statements related to classroom practices in teaching arithmetic to students in their third year of elementary school. The acceptance of the Q-sorting task by the teachers in both the graded and non-graded groups was very good. A typical response was that of one teacher who commented, "I really analyzed these cards and rather enjoyed doing this sorting--found it not too simple to make a few decisions on a few of the cards."

As to the ambiguous or vague question, only six teachers listed statement numbers. The total number of statements were 13. Two teachers agreed on only six statements as being vague. The remaining seven statements were isolated selections. The approximate sorting time was an hour for most of the teachers. The sorting time range was from 30 minutes to three hours.

The Q-sort matrices for both graded and non-graded classroom teachers are tabulated in Table 1. Under each statement number is the pile number, 1 through 7, in which each of the 70 cards was placed as the teacher sorted. The sums of the ranks for each card were considered representative of the ten teachers in each group. For example, the sum for card one in the grade teacher sort was 38. The average rank for card one was 3.8. Schill states, "The resultant sort, which is arrived at by averaging the responses to each item for all individuals who responded, can be used as a consensus of opinions solicited."[4]

Summary

This study was designed to analyze and compare the individualization of arithmetic instruction in graded and non-graded elementary schools in selected districts in Oregon. Instruments and procedures were developed to test the following hypotheses, which formed the bases for this investigation:

1. There are no significant differences in the way graded and non-graded classroom teachers conceptualize the individualization of arithmetic instruction.
2. There are no significant differences in the

individualization of instructional practices in arithmetic within selected graded classrooms or within selected non-graded classrooms.

3.   There are no significant differences in the individualization of instructional practices in arithmetic between selected graded and non-graded classrooms.

The data gathering procedures included (1) Q-sorting by each of the 20 teachers.   The seventy statements, on cards, described classroom instructional practices in arithmetic.   (2) Observations were made in each classroom during an arithmetic session and an interview was held with each teacher.   All teachers were rated on 21 items which described individualization of arithmetic instruction.   (3) The student variability in each classroom as measured by the Contemporary Mathematics Test was assessed.

Analysis of the Data

1.   Kendall's Rank Correlation Coefficient was used to interpret the agreement in Q-sorting between the graded teacher group and the non-graded teacher group.   A correlation coefficient of .73 was computed and found to be significant at the .01 level.   Therefore hypothesis number one was accepted.   There were no significant differences in the way teachers in graded and non-graded elementary classrooms conceptualized the individualization of arithmetic instruction.

2.   Using an observation-interview check sheet as the data source, Chi Square was applied to determine the differences within and between the two groups of teachers.   Each group of teachers differed significantly in nine of 21 practices listed on the check sheet.   On the basis of this analysis, hypothesis two was rejected.   There were significant differences in the individualization of instructional practices in arithmetic within graded and non-graded classrooms.

3.   The between group comparison on the same 21 items showed that the 20 teachers differed significantly on only one item.   Consequently, hypothesis three was accepted. There were no significant differences in the individualization of instruction in arithmetic practiced by teachers in graded and non-graded elementary schools.

TABLE 1. Q-Sort CARD RANKING-TEACHERS IN GRADED SCHOOLS FOR FIRST 35 STATEMENTS

Teacher	Numbered Statements																
	1	2	3	4	5	6	7	8	9	10	11	12	13	14	15	16	17
G1	1	2	5	4	3	5	3	3	4	7	4	3	1	6	4	6	6
G2	6	2	7	3	6	5	3	4	2	7	7	2	1	3	6	7	3
G3	5	5	3	2	3	2	6	6	3	4	4	2	2	6	7	7	2
G4	1	7	4	4	5	4	1	6	4	1	7	2	2	3	4	7	4
G5	3	5	3	2	4	2	5	5	2	4	7	2	1	7	6	2	6
G6	4	2	6	5	5	3	2	5	3	4	6	1	2	2	7	5	4
G7	5	2	7	1	5	4	2	4	2	7	7	1	1	3	5	4	6
G8	5	3	1	3	7	5	2	4	2	6	7	3	2	6	1	7	6
G9	6	3	5	2	7	4	3	5	4	7	7	2	2	1	5	5	5
G10	2	3	6	3	6	5	4	6	2	4	5	2	1	3	5	2	6
Rank Sums	38	34	47	29	51	39	31	48	28	51	61	20	15	40	50	49	48

Teacher	18	19	20	21	22	23	24	25	26	27	28	29	30	31	32	33	34	35
G1	2	2	1	2	3	1	1	5	1	1	5	5	6	1	6	2	7	7
G2	1	2	1	5	1	2	1	4	3	1	4	4	1	2	6	7	3	6
G3	3	7	5	5	1	1	1	3	1	6	3	2	6	1	6	7	7	4
G4	5	3	5	3	2	4	1	1	5	1	4	3	1	4	3	5	7	5
G5	2	4	2	3	1	4	2	5	1	3	1	3	3	1	7	5	7	5
G6	1	2	1	6	1	2	1	1	1	5	4	5	3	1	5	7	7	6
G7	3	3	4	5	1	2	1	3	3	3	2	2	4	1	6	4	5	6
G8	1	1	2	4	1	4	5	3	7	3	6	1	4	2	4	2	6	7
G9	1	6	1	5	1	2	3	5	2	2	6	4	2	4	3	6	6	3
G10	2	2	3	1	1	3	1	4	2	4	1	3	6	1	7	3	7	6
Rank Sums	21	32	25	39	12	25	17	34	26	29	36	32	36	18	53	48	62	55

Conclusions

Given two randomly selected groups of teachers, one group teaching in graded elementary schools and the other teaching in non-graded elementary schools, the differences were not significant between the two groups in their choices of arithmetic practices, which were considered along a continuum from least characteristic to most characteristic of individualizing instruction.

Though Q-sorting has not been widely used in curriculum studies, it is gaining in use and in professional acceptability.   To ferret the information needed to test the first hypothesis, a unique and creative instrument which was interesting, challenging, and easy to use was required.   No other instrument met these criteria.   Rejection or acceptance of this first conclusion may hinge on replication or other validation procedures related to this particular Q-sort approach.

In examining actual teaching practices some differences did exist within both groups when checked against items which were based on individualizing practices suggested in the literature and which were examined to be reasonable in the light of this writer's experience.   It seems justifiable to conclude that this variability is dependent on other criteria than a graded or non-graded label.

Notes

1.   Lloyd H. Silverman.   "A Q-sort Study of the Validity of Evaluations Made From Projective Techniques," Psychological Monographs 73(7):1-28; p. 2.
2.   Stanley A. Perkins.   A Comparative Analysis of the Congruence of High Creative and Low Creative High School Students.   Doctoral dissertation.   Eugene: University of Oregon, 1966; p. 54.
3.   Fred N. Kerlinger.   Foundations of Behavioral Research. New York: Holt, Rinehart, and Winston, 1964; p. 598.
4.   William J. Schill.   "The Use of Q-Technique in Determining Curriculum Content."   California Journal of Educational Research 12:182, September, 1961.

MERITS OF SELECTED ASPECTS OF
INDIVIDUALIZED INSTRUCTION

Victor L. Fisher, Jr.

The essence of planning a sound elementary school
mathematics program to meet the needs of the individual student grows out of an understanding of how the psychological
traits of individuals differ.   Only when educators become
more knowledgeable of the extent to which students differ,
and more capable in adjusting methodology and curriculum
content to meet these differences, can the elementary school
mathematics program approach a uniquely individual program
for each student.

The purpose of this research is to study selective
facets of an "individualized" mathematics program for elementary school pupils.   More specifically this study will attempt to ascertain what happens to students' achievement
when they work and progress independently in their study of
elementary school mathematics, and what happens to students' achievement when they check the answers to their
problems as they work through an assignment.

Definition of Terms

INDIVIDUALIZED MATHEMATICS PROGRAM: one
that is tailormade to fit the needs of the individual student.
It must include the following:

(1)   Each student receives instruction at his own level
of achievement.
(2)   A wide range of instructional materials is available in the classroom.
(3)   Each child is permitted to progress at his own
rate.
(4)   Each child is permitted to meet with the teacher
individually or in small groups of students with similar problems.
(5)   Each child is permitted to check his own work as
he works through the assignments.
(6)   The slow learner is not required to meet the
standards of the group and the bright child is permitted to

explore areas of mathematics in which he is interested.

CONVENTIONAL MATHEMATICS PROGRAM: one that has the following characteristics:

(1) Instruction is given to the entire class at one time and is in the form of lectures, demonstrations, and/or discussions.

(2) All students receive instruction from the same textbook.

(3) No differentiation is made in assignments to meet individual needs or abilities.

(4) All students are expected to progress through the development of a concept at the same rate and to cover a set amount of material in a given year.

(5) The student's daily work is checked by the teacher and returned one or two days after the student has completed it.

(6) All students, regardless of mathematical ability, compete against one another for grades within a predetermined set of standards.

The Sample

The sample for this study was selected from the 120 students enrolled in the sixth grade during the 1965-1966 academic year in an elementary school located in a large urban community in a midwestern state. The students were systematically assigned to four classes at the end of the 1964-1965 school year. This was accomplished by listing the students alphabetically and then assigning every fourth student to the same group.

There were two experimental groups, each receiving a different type of treatment, and one control group. Three of the four sixth-grade groups were randomly selected to participate in the study. Random assignment was accomplished by listing the groups by room numbers and then utilizing a table of random numbers.

The program designed for experimental group A, experimental group B, and the control group C had several facets which were common to all groups.

(1) Each group used the basic textbook, Elementary School Mathematics, Book 6. This textbook was selected for

three reasons:
(a)   The authors include recognized educators and mathematicians.
(b)   The subject content of <u>Elementary School Mathematics, Book 6</u>, is considered by educators to be "modern."
(c)   The textbook was the adopted textbook for the sixth grade of the school corporation in which this study was conducted.   Therefore, the continuity and sequence of content established by the authors in the series for grades one through six was not disturbed.

(2)   Work assignments for the three groups were identical. The assignments were determined by a study of the daily mathematics assignments made by the four sixth-grade teachers of the participating elementary school during the two years prior to this study.
(3)   Each class was conducted in the same room in order that the same mathematical aids would be available to all students.
(4)   All students experienced the same evaluation procedures.
(5)   All work was done within a 40-minute class period.

In experimental group A the students were permitted to work and progress independently in their study of mathematics.   They progressed through developed assignments which were identical for all groups, both experimental and control.   The students were guided in their work by an assignment sheet which listed the problems they were to work in each section.

LESSON ASSIGNMENT SHEET GRADE 6

Lesson 20
Page 44; Study the examples; Work exercise 3;
Page 45; Work exercise 1
Lesson 21
Page 46; Study the examples; Work exercise 3
Lesson 22
Page 49; Work exercises 5-7 at top of page;
Work exercises 1-10 at bottom of page
Lesson 23
Page 50; Study the examples; Work exercise 1
Lesson 24
Test 5
Lesson 25
Page 51; Work exercises 4 and 5

Lesson 26
———————
Page 52; Study and work the examples; Work
exercise 1

Lesson 27
———————
Page 52; Work exercise 2

Lesson 28
———————
Page 53; Work exercise 4

Lesson 29
———————
Test 6

Lesson 30
———————
Page 53; Work exercise 5

Instruction was given by the teacher to the individual students or to small groups as they requested it, or needed it (as indicated by their problem sheets), or as they approached a new skill. Each student had available an answer sheet. When a student completed a problem or an assignment he checked it with the answer sheet, and the problems which were incorrectly worked were reworked. The work sheet was given to the teacher for further evaluation.

Each student in experimental group A who had a need for additional work on a particular concept either worked the problems in the textbook that were not used as a part of the regular assignment or utilized the "Supplementary Exercises" in the back of the textbook. If the teacher believed a student needed additional work, assignments were made from these sections.

Experimental group B received instruction as a group and progressed from topic to topic as a group. Assignments made daily to the entire class by the teacher were identical to those used for group A. It should be noted, however, that students within this group were not permitted to advance to the next assignment upon completion of an assignment, as group A was permitted to do. They were instructed to begin an assignment only when the teacher had made the assignment to the total group. Also, as in group A, each student in group B had an answer sheet available and, as he completed a problem or an assignment, he checked it himself. Problems which were incorrectly worked were reworked and the work sheet was given to the teacher for further evaluation.

When the teacher's evaluation of his class indicated that the majority of the class were not understanding a concept, he made an additional assignment from the problems

not utilized in the textbook for the original assignment or
from the "Supplementary Exercises" section in the textbook.
The additional assignment was made to the entire class.

Control group C received instruction as a group and
progressed from topic to topic as a group.   Assignments
were made daily by the teacher to the entire class.   There
was no differentiation in assignment to adjust for individual
differences.   The assignments were identical to those used
for groups A and B.   However, students within this group
were not permitted to advance to the next assignment upon
completion of an assignment, as group A was permitted to
do.   They were instructed to begin an assignment only when
the teacher had made the assignment to the total group, as
was done in group B.   When a student completed an assignment he gave it to the teacher for evaluation.   The teacher
evaluated each student's daily work sheet and returned it to
him the next school day.   Problems which were incorrectly
worked were reworked and returned to the teacher for further
evaluation.

Remedial work for the control group was conducted in
the same manner as for experimental group B.   When the
teacher's evaluation of his class indicated that the majority
of the class was not understanding a concept, he made an
additional assignment from the problems that were not utilized in the textbook for the original assignment or from the
"Supplementary Exercises" in the textbook.   The additional
assignment was made to the entire class.

Every fifth assignment for experimental group A, or
once each week for experimental group B and control group
C, a review lesson, covering the preceding four assignments,
was given.   This was placed in the treatment first, to give
the teacher a means of evaluating the student's progress;
second, to provide additional evidence for assigning grades
which were required by the school corporation every six
weeks; and third, to attempt to reduce the probability of mis-
use of the answer sheets by the students in the experimental
groups.

The study included 92 assignments.   This covered the
mathematical concepts recommended for one semester's work
in the sixth grade by the school corporation in which the
classes participating in the study were located.

Summary

 The comparison of achievement in mathematics reasoning and mathematics fundamentals, as measured by the arithmetic section of the California Achievement Test and information obtained from student cumulative record folders, provided the basis for determining statistically the contribution two facets of an individualized mathematics program made to students' achievement in elementary school mathematics.

 Within the scope of this study and for the population tested the following findings were discernible:

 (1) There was no significant difference among the three treatments regarding the students' achievement in mathematics reasoning or in mathematics fundamentals.

 (2) There was a significant difference in achievement in mathematics reasoning and mathematics fundamentals between the students in the low socio-economic group and the students in the average socio-economic group. The students in the average socio-economic group had a higher mean score on the arithmetic reasoning test and on the mathematics fundamentals test than did the students in the low socio-economic group.

 (3) The students in the different socio-economic groups did not differ significantly in achievement in mathematics fundamentals or reasoning among the three treatments.

 (4) There was no significant difference in achievement in mathematics reasoning or in mathematics fundamentals between the sexes.

 (5) The sexes did not differ significantly in achievement in mathematics reasoning or fundamentals among the three treatments.

 (6) The majority of the students in experimental group A who were permitted to work and progress independently in their study of elementary school mathematics completed and the required work in less time than the students taught by the conventional group method. In addition to the reduction in time, the students in experimental group A achieved a higher (though not a statistically significantly higher) mean score on the post-test than did the students taught by the conventional group method in experimental group

B and control group C.

(7)   The students in the two experimental groups who were permitted to check their work as they completed each problem achieved a higher (though not a statistically significantly higher) mean score on the post-test than did the students who had their work evaluated by the teacher and returned to them the next school day.

INDIVIDUALIZED TEACHING OF FIFTH- AND SIXTH-GRADE ARITHMETIC

William A. Graham

For the past several years much thought has been given to development of techniques for individualized instruction in the arithmetic skills at the Kilby School; in reading, all the teachers now use individualized teaching.   The results have been gratifying and the children are enthusiastic and cooperative.   During 1959-60 the fifth- and sixth-grade teachers decided to teach arithmetic on an individualized basis for their entire class.   This report will try to tell something of how the work was done and a few outcomes as shown by test results and opinions of the teachers.

Fifth Grade

The class were tested by the Iowa Every Pupil Test of Basic Skills in November.   The range in arithmetic scores was from 4. 4 to 6. 3 with a median of 5. 5.   They were also given the California Test for Mental Maturity, and the range in IQ scores on this test was from 102 to 148 with a median of 117.

The children were further tested by the teacher to see which skills they had mastered and each one was then started to work on the process in which he needed help.   If several children needed the same help they were grouped for teaching the meaning of the process with concrete aids.   After this they were given assignment sheets covering practice in the work taught.   These sheets were checked.   When the process seemed to have been understood, the child went on to a new

lesson.   This sequence within processes was carefully
planned.   For instance in division these steps were taken.

(1)   Basic division facts, as 42 ÷ 6.
(2)   Division with a remainder, as 46 ÷ 7.
(3)   Two figure quotients, as 126 ÷ 6.
(4)   Three figure quotients, as 684 ÷ 6.
(5)   Two figure division, as 20 ÷ 10, with increasing
difficulty up to such ones as 421 ÷ 29.
(6)   Checking of division.

The work materials were kept in a place accessible
to the children and they were able to take care of their own
needs.   At arithmetic work time the children worked about
40 minutes at their own speed on the individual assignments.
If two or three were working on the same thing they tended
to group themselves and work along together.   This was encouraged since they were able to help one another.

The units completed were checked by the children
and corrected before they were placed in the individual files.
Tests were checked by the teacher and corrected by the
children before filing.   The class had access to three current fifth-grade texts and when the teacher felt that one of
these presented a topic better than the others it was used
by all needing help in that topic.   Another form of the Iowa
Test was given in May.   At that time the range was from
4. 6 to 8. 7 with the median at 6. 7.

Sixth Grade

The sixth-grade teacher defined individualized arithmetic teaching as a procedure in which children are grouped
according to their needs and are then permitted to move
along at each child's own rate.

At the beginning of the year a checkup was given.
This test covered the fundamental processes.   Using the
test results as a guide, the children were grouped according
to needs.   If one appeared to need only a short review he
was moved on quickly, but if there was any lack of understanding of a process it was carefully explained, and as
much practice as any pupil needed was allowed him.   Children who could use the fundamental processes worked on
solving problems in area, volume, and the metric system,
planning trips, making charts and graphs, and solving a

variety of written problems.

A test was given every two weeks as a review of the fundamental processes and to determine the needs of individuals for grouping.   Materials used consisted of arithmetic texts, mimeographed sheets, and problems from social studies and science.

Since groups were organized according to needs, individual assignments were made on that basis.   When a child completed an assignment or thought he understood a process he was checked and if the test showed the desired mastery, he would move on to another step or continue practice in the same work if such work was indicated.   Problems using the processes taught were in continuous use.

This class was tested with one form of the Iowa Every Pupil Test of Basic Skills in arithmetic in November and with another form in May.   In November the scores ranged from 5.0 to 9.4 with the median of 6.4.   In May the range was from 5.9 to 11.3 with the median at 8.0. According to the California Test of Mental Maturity, the IQ scores for the class ranged from 87 to 146 with the median at 115.

The teachers felt that the most evident outcome was heightened interest in mathematics.   Other outcomes were independence in working with much less reliance upon the teacher and the wide divergence of rates for growth.   Some children made as much as two and a half years' progress during the interval between the tests.   The children were more heterogeneous at the end of the year than at the beginning.   Care had to be taken to see that the very dependent child did not get confused, but as the program developed these pupils tended to gather themselves into small group situations.   The teacher then worked with them and their problems.   Children found security in being with others like them in their class.   It was interesting to see the children electing to do arithmetic during their free-time periods.

AN EXPERIMENT IN INDIVIDUALIZED ARITHMETIC

Caroline C. Potamkin

"Jerry and I are ready for two-place multipliers. "
"I need a lesson in finding averages. "
"I am up to subtraction of four-place numbers and so
is Ellen. "

Remarks like these may be heard in our fourth-grade
classroom during a typical arithmetic period. They are not
indications of a state of utter confusion; they are the outcome
of a planned system of individualized instruction, which has
been evolving gradually over the past several years. It
originated very simply with an attempt to have each child
work a limited sequence of textbook assignments at his own
rate of speed.

We had spent considerable time on review and diagnostic work, using charts and manipulative materials to
establish basic arithmetic concepts, and were ready to go
ahead in ability groups. What would happen, I wondered, if
the children, instead of being grouped at this point, were
allowed to start on textbook work independently? What would
happen if each child went on to the next assignment as soon
as he had correctly completed the first one?

With the wide range of ability in any average class,
I knew that some children would complete three assignments
while others were still working on the first, and eventually,
I thought, the groups would establish themselves automatically. But I was mistaken. We developed in an entirely different direction.

We started, on Monday, with this assignment on the
board: PAGE 32, EXAMPLES 1-10. "When you have finished, " I told the class, "bring your papers to me for
checking. "

As soon as the first child had completed the assignment correctly, I placed a second assignment on the board:
PAGE 34, TOP OF PAGE, ROWS 1-6. A few children had
finished two assignments, and before the arithmetic period
was over I had listed a third. Many were still working on

the first.   This meant that on Tuesday I had to list all the
assignments from the beginning, to take care of the slower
children, and keep adding new assignments for the more
able learners.

It seemed more efficient to ditto a list of assignments
from the beginning of this unit of work, and give a copy to
each child.   The first three items on the list read: P. 32,
examples 1-10; P. 34, top of page, rows 1-6; P. 35, problems 1-8.   Similar items followed.   I listed only the textbook material that I felt the children needed in working on
this particular concept.   I still hoped to set up a fast group
made up of the children who finished the list first.

I told the class that this ditto copy was their worksheet; each of them was to follow it independently; each step
must be checked and corrected before the next step could be
attempted.   The children enjoyed this new way of working
and the feeling of independence that went with it, but I found
myself swamped with the task of checking papers   It became
impossible to check each child's work as soon as he had
finished, but it was also impossible to let a child go on unless he had finished his work correctly.   Thus the device of
having the children check their own papers was forced on
me.

Making up one set of answers for each assignment
was not satisfactory, since several children were usually
ready at the same time.   Therefore, I ran off ten copies of
the answers for each of the assignments listed on the worksheet, clipped each set of ten copies together, and placed
them on a shelf in manila folders.   On the front of each
folder, using a felt pen, I printed:

ANSWER SHEET P. 32, EXAMPLES 1-10 and
ANSWER SHEET P. 34, TOP OF PAGE, ROWS 1-6

and so on.   Several sets could be kept in one folder.   The
following instructions were printed on a chart, which was
mounted on the bulletin board:

Use the answer sheets to check each assignment.   If
you made mistakes, turn over the answer sheet, correct your work, and check again.   When your work is
correct, put the answer sheet back, put your paper in
the folder marked "Finished Work, " and go on to the
next assignment.

In this way I could go over the papers quickly at another
time, without causing delays.

This plan worked in more ways than I had foreseen.
As I had hoped, I was relieved of the frenzied attempt to
keep things moving, and I was able to walk around the classroom and check on the children who were having difficulties.
(Two advantages I had not planned soon became apparent.
First, the children were able to learn the results of their
efforts immediately. The information was a highly effective
form of motivation. Second, the children could discover
their own successes and failures while interest and retention
were at their peak.)

A few of the children were now nearing the end of
the assignments listed on the worksheets, and I began to
have second thoughts about grouping. Since the children enjoyed work independently, and interest and motivation were
high, why not continue with a second worksheet?

This extension of the plan brought up two major ob-
stacles. As the children came to new steps in the arithmetic process, they would need instruction, and their work
would have to be evaluated. Still in an experimental frame
of mind, I made up Worksheet 2, modifying the format in
an attempt to meet these problems. The sheet looked something like this:

Worksheet

New Lesson: Subtracting four-place numbers
P. 50, examples 1-8
P. 52, top of page, examples 1-10
P. 53, problems 1-8
P. 54, examples 1-8
Test, P. 55, whole page
New Lesson: Uneven division

When the children came to the entry "New Lesson"
on the worksheet, they were instructed to inform me, and
I would call them up to the blackboard for instruction.
Enough copies of this second worksheet were dittoed so that
each child could have one. Copies were placed in a folder
so that each pupil could take one whenever he was ready.

At this point, I had no idea as to how these innovations would work. I thought I might find myself swamped

with children clamoring for instruction on several processes
at the same time, as I had been swamped with checking papers.   As it turned out, these problems did not arise.
Three children were ready for the new lesson on subtraction
while the others were still working on the first worksheet.
By the time these children were ready for the lesson on uneven division, I had already worked with several small
groups (anywhere from two to five, usually) on the subtraction lesson.   There was enough spread in rates of work that
I was able to keep up with the groups as they came up.

While it is true that I was teaching the same subject
matter to different children at different times, this was not,
I felt, an inefficient use of time.   The lesson was presented
to each child as he became ready for it.   Moreover, I was
able to present the work in varied ways, depending on the
needs and the abilities of the child and, in working with very
small groups, I could diagnose and correct errors and mis-
understandings before incorrect habits became fixed.   As a
not inconsiderable bonus, I was in the enviable position (for
a teacher) of having children come up and ask me to teach
them something.

As the children progressed, the spread between the
fast and the slow learners became more pronounced.   In the
same arithmetic period I might be teaching the first new
lesson to one child, and the fourth or fifth new lesson to
another.   The advantage of the plan was that lessons could
be taught out of sequence without upsetting the program in
the least.

I had solved the problem of answer sheets when the
children were working on the original worksheet.   Now I
ran off answer sheets for the assignments in Worksheet 2.
There were no answer sheets for the tests.   These were
placed in a marked folder, and since they were completed
at widely varying times, I was able to check them quickly,
record the grades, and return them.   When I returned a
test, I gave the pupil permission to continue with the next
step on the worksheet.   Or, if necessary, I gave instructions for review.

Since the children were working independently, they
sometimes needed to be helped over difficulties.   I could
devote some time to this assistance, but because much of
my time was taken up with teaching new work and checking
tests, I was not always available when the children needed

help. To solve this problem, I appointed two of the more able children "teachers for the day." Anyone who needed help was to consult them. These "teachers" were changed each day, so that no one child had to devote a disproportionate amount of time to helping others. The actual process of helping was valuable in itself, however. It not only encouraged maturity and social responsibility, but it developed deeper insights in arithmetic thinking, since the "teachers" had to find out what the other child was doing wrong and how best to help him.

By the time the inevitable frontrunners were nearing the end of Worksheet 2, I had decided that our experiment in the individualization of arithmetic was too successful to be abandoned and prepared Worksheet 3. Following the system of new lessons, assignments, and tests, I made another modification. About half way down the sheet I listed:

New Lesson: Equation
Equation Sheet 1

There was no equation work in our textbook. Since I wanted to introduce this concept, I dittoed a sheet of equations that I had made up. After this new lesson, the children had to find the folder labeled "Equation Sheet 1" and work from it. Interspersed among the assignments on subsequent worksheets were equation sheets 2, 3, and so on, with answer sheets filed and labeled separately. These were of graduated difficulty.

At the bottom of this third worksheet, I listed:

Extra Credit--Special Worksheet 1

This was an enrichment exercise, aimed at challenging the more able learners and not part of the required work. In later worksheets, there were more assignments for extra credit.

On the next worksheet, I began to list review exercises. These were taken from the textbook and labeled "Review exercises for those who need review. Ask the teacher."

As time went on and I felt the children were ready, I added more work that was not in the textbook. The children understood that these special assignments were not in the

textbook but were filed in marked folders.   As they progressed from one assignment to the next, and from one worksheet to the next, they became adept at locating the ditto
sheet they needed.

It was essential that the material be organized, clearly labeled, and accessible.   We spread out the various folders on open shelves.   Several sets of worksheets or answer
sheets could be kept in one folder, if the contents were
clearly marked on the front.   A committee of two children
was assigned to check the folders, to make sure that all
dittos had been replaced in the proper folder and clipped together in the same set.   At the beginning of the program,
the material was reorganized at the end of each arithmetic
period.   But it soon became necessary to postpone the job
until the end of the school day; the children were using the
folders throughout the day, whenever they had spare time.

While we were now committed to an individualized
program, and continued to carry it on, we still had occasions
for classwork.   Time was set aside for drill in the number
combinations, and occasional periods were scheduled in such
areas as the study of measurements and the manipulation of
fraction parts.

Advantages of Answer Sheets

Many questions have been asked about the use of answer sheets by children.   The questions all boil down to one
concern: cheating.   In my experience with several fourth-
grade classes, cheating has not been a problem.   The atmosphere is not conducive to cheating.   The children are eager
to see how well they have done.   They take pride in being
successful and in finding and correcting their own errors.
There is no stigma attached to asking for help or to working
more slowly than the child in the next seat.   Then there is
always a test coming along in the near future, in which any
basic failures in comprehension will be discovered the hard
way.

The benefits of the use of answer sheets, as I indicated earlier, far outweigh any possible drawbacks.   Besides
freeing the teacher for work with individual children, the
answer sheets help to fix learnings, because they are used
at the moment when the child is most deeply involved in a
particular piece of work.   For this reason, and because they

offer immediate rewards, they provide strong motivation.
"I want to finish and see how I did," the children say. The
purposeful manner of the children as they go to the designated shelf, find their answer sheet, and take it back to
their desks for checking, is revealing.

Planning the Program

Another question is frequently asked: "How much time
is required to start this kind of program and to keep it going?" The task of planning seems formidable when the program is viewed in its entirety. In actual operation, however, it is a step-by-step process. All that is needed to
start is Worksheet 1 with its accompanying answer sheets.
If the teacher follows the curriculum and the arithmetic textbook, this first worksheet is not difficult. Subsequent worksheets, and additional material from sources other than textbooks, can be developed as the children progress.

Working in this way, of course, eliminates the chore
of daily lesson plans; and while the arithmetic period is a
busy time for the teacher, it is the busyness of implementing an activity that carries its own momentum, rather than
the busyness of initiating a new activity each day.

As in any teaching situation, the degree of success
is reflected in the reactions of the class. I found that my
fourth-graders developed an interest in arithmetic. They
were using their spare time throughout the day to do their
worksheet assignments or to check answers. They looked
forward to arithmetic time, and, as they became involved
in the program, they became self-starters. Books and
worksheets came out without a word from anyone. Work
began. Everyone knew what to do.

One Arithmetic Period

On a typical day, I began by assigning the "teachers
for the day." I wrote their names on the board. I then
asked who was ready for new work. Ann, David, Bobby,
and Sara were ready for division with remainders. Allan
said that he had had the lesson yesterday, but he would like
some more work on it. Could he come up, too? After a
few minutes Ann, Allan, and Bobby were sent back to their
seats ready to work on their own. I worked with David and

Sara for a while longer.

Susan, one of the "teachers," came up to tell me
that Jack was having trouble because he did not know his
six tables. Jack was set to working with the flash cards
before going on with his worksheet assignment. There was
no question of his falling behind; he would simply pick up
where he left off. Sidney had finished a test, which I
checked quickly. He had done well and could go on. Mary
and Frances were ready for equations--and so on.

Meanwhile, everyone was working, checking, or consulting the "teachers." There was a certain amount of
movement and talking. Children were going back and forth
to pick up or to return ditto sheets; those who needed help
were talking to the "teachers"; there was talking in the front
of the room, where I was working with a group. However,
it was a working noise that arose from the nature of the
activity. There was no discipline problem--and this in a
class of 32 children, with at least its share of so-called
discipline problems.

Jimmie, a notoriously intractable child and a slow
learner, was working busily, pleased to find that he could
complete an assignment correctly (after much help and encouragement). True, he was still way back at the beginning
of the first worksheet, but he was making his own kind of
progress and enjoying the satisfaction that went with it.
Arlene, a daydreamer, was more difficult. She was not
working fast enough for her capacity and had to be checked
on. But even Arlene was getting the feel of working on her
own and being responsible for herself. In short, this plan
of operation is not in the realm of theory or conjecture. It
originated in an actual classroom situation, and it works.

As it evolved, this plan for individualizing arithmetic
came to resemble (quite inadvertently) those aspects of programmed learning that are most conducive to the child's development. As in programmed learning, the children follow
a planned sequence at their own pace and are apprised of
the results of their work as they go along. As an invaluable
by-product, they are acquiring independent work habits and
a sense of responsibility.

But in our system the material is not self-teaching.
It does not lead inherently from one correct response to
another. Instruction and correction, according to individual

needs, are basic. This system is a combination of traditional and programmed instruction, planned, controlled, and
administered by the teacher to meet the needs of the children.

While this plan has been confined to arithmetic, the
method can be used with other subjects. Once the children
have adjusted to this type of independence, they can readily
transfer these same work habits to spelling or social studies,
for example. The actual planning would differ, and the
methods could be modified in any way dictated by the subject
matter or the preferences of the teacher, but the principle
would remain the same.

As it happened, this plan was used in a fourth-grade
class. But it could have been used in any elementary-
school grade from third through sixth. Furthermore, since
the content is flexible, and can consist of material of much
greater complexity, it can be applied to secondary schools
as well. There is no reason why students in junior and
senior high school could not benefit from the flexibility, the
independence, and the immediate confirmation of results that
this plan offers.

This seems a great deal to claim for a technique that
consists of a few simple steps: listing a sequence of assignments on worksheets, providing answer sheets, teaching
new work as the children become ready for it, providing
helpers from among their own classmates, varying the content to provide enrichment and adaptability, and checking
tests. However, basic changes are involved--changes in the
way children work, in their relationship to one another and
to the teacher, and in the role of the teacher.

The Effect on Pupils

Children feel very differently about what they are doing when they themselves are in control. They are, so to
speak, in business for themselves and responsible to themselves. This freedom of thought and action provides an incentive at any age. The experience is maturing, but not
threatening, because help is available whenever it is needed.
The knowledge that success can be ascertained immediately
inspires further efforts. Failure is not disastrous, since
there is always another chance and plenty of support. Self-
esteem is enhanced, because each child is being accorded

the respect due him as a special person, with his own ways
of working and his own strengths and weaknesses.

By helping each other, the children develop an interest in one another's progress and a real satisfaction in helping another child to succeed. Empathy is encouraged, and
an atmosphere of co-operation prevails. No one is being
threatened by another child's success or held back by someone else's failure.

The Teacher as Helper

In relation to the teacher, the children have the benefit of greater personal contact. They have her full attention
when she works with them in small groups. She knows what
their problems are and works directly with each child to
solve them. She is not a taskmistress, but a person who
is there to help.

The role of the teacher, then, is that of a helper.
It is her function to teach, but only when asked. She does
not visibly direct the work. She simply expedites it. She
is, in a way, the catalyst, setting things in motion, freeing
the children to move ahead under their own steam. Instead
of being circumscribed and hemmed in by the indoctrination
of prescribed doses of knowledge, the pupils are given a
direction and released to follow it.

Intellectual energy is present in every normal child,
but it is not easily drawn out. If we can find the key that
releases this energy along intellectual lines, we have found
the key to growth. The significance of this experiment lies
in the attempt--not to coax, cajole, or even stimulate--but
to release children's energies in the direction we would like
them to follow.

VII. RESEARCH: 1. IPI

In this chapter and the next we return to the IPI program. The very excellent description of the program that is contained in Dr. Scanlon's thesis, from which the first article has been extracted, has been omitted, as I have had two very adequate descriptions of IPI in the previous chapter devoted to that program.

Robert G. Scanlon

Dr. Scanlon is presently program director of the Individualized Learning Program of Research for Better Schools, of Philadelphia. In that capacity, he has jurisdiction over a considerable number of IPI demonstration schools, many of them within a hundred miles of Philadelphia. In his thesis, which was completed at the University of Pittsburgh in 1967, the author sought to ascertain to what degree the kind of individualization involved in IPI procedures would lead to self-initiated classroom activities by pupils in the fifth and sixth grades. The portions of this thesis which I selected for inclusion in this chapter are of uncommonly high interest as well as significant for my thinking about individualization of instruction. I commend this material to the reader for careful and thoughtful examination.

John L. Yeager

Dr. Yeager is associate director of the Learning Research and Development Center at the University of Pittsburgh. As his doctoral project at the same institution, which he completed in 1966, Dr. Yeager investigated the rate of learning of first- through sixth-grade children who were being exposed to the IPI approach.

If one stops to think about it, a program should

certainly not be measured solely on the basis of the amount
learned, but also on the efficiency with which that learning
has taken place.   One measure of the degree of efficiency
might well be the rate at which learning takes place.   The
author of the thesis from which the material in the second
part of this chapter is taken is the first to call attention to
the fact that learning rate is an extremely difficult item to
single out in pure form.   For example, if I spend an hour
learning a task, there is no way that I know of for anyone
to determine how much of that time I really spent in the
learning process and how much time I may have spent with
a frown of concentration on my face, thinking and day-dream-
ing about an upcoming vacation.   I once said to an aged
aunt: "Aunt Mary, a penny for your thoughts. "   Her reply
was, "My thoughts are my own. "   This is not to detract in
any way from the importance and merits of a genuine effort,
such as we find here, to reach findings that are as nearly
valid as possible within the scope of our present knowledge
about such things.

I think that one other caution is in order in reading
the Yeager selection.   It must be remembered that the only
thing that a statistically significant correlation co-efficient
can tell us is that it differs from zero.   This is a point that
is often overlooked.   Actually, a correlation coefficient that
is statistically significant does not have a very high standing
in this regard.

One of the advantages of large scale research, such
as a program like IPI, is that it produces research findings
that can shed light not only on that particular program but
also on comparable, or for that matter, contrasting programs.   In my opinion, the theses written by Drs. Scanlon
and Yeager represent a very high degree of excellence in
such research.

Donald Deep

Dr. Donald Deep, who completed his doctoral work
at the University of Pittsburgh in 1966, chose as the topic
for investigation to be reported on in his thesis the assessment of the impact of a program of this type on pupils of
different levels of mental ability.

This is an extremely significant topic because there
seem to be such strong convictions on this subject by both

those who have and those who have not made an investigation
of the matter.   There is one school of thought that states
with great assurance that any program of individualization
should only be used in teaching children who possess superi-
or mental endowments.   Proponents of this theory appear to
reason that in some indefinable manner any sort of individualized procedure must be more difficult than the traditional
approach and that therefore only the mentally superior are
likely to profit from it.   This group is equally certain that
individual differences have their sharpest impact on pupils
who are able to proceed more rapidly than their classmates.

Another school of thought adopts the "don't rock the
boat" approach and reasons that only those of less than average mental capacity are likely to profit from individualized
instruction.   The argument runs that those who are already
doing well should be left alone as quarrelling with success
is certainly tempting the fates.

There are two matters that one should consider in
evaluating these two points of view: do these arguments
concern themselves with ability, as they say they do, or
are they actually concerned with differing levels of achievement? and is there any assurance, merely on the basis of
the fact that certain classes of pupils are doing well, that
they are therefore working under optimum conditions?   Is
it not possible that their achievement might be even further
enhanced by the use of a different approach?   It was Dr.
Deep's purpose to gather factual evidence concerning these
matters.

The reported results were based on experimental procedures at three schools.   It is seen in the report on the
findings of this study that there did not appear to be any
great differences in the effect of IPI on children of varying
mental abilities in the fourth, fifth, and sixth grades.   If
these findings are accepted as valid, then it appears that the
decision concerning the use of individualized instruction
should be based on considerations other than the mental ability levels of the pupils.

A PROGRAM FOR ENCOURAGING SELF-INITIATED ACTIVITIES

Robert G. Scanlon

I.   Introduction and Related Research

A major challenge that has faced education is how to provide for the differences in pupil aptitude and interest within the framework of a school program necessarily geared to mass education.   Many American schools have placed an emphasis upon making provisions for individual differences and developing programs of individualized instruction.   Recently, interest in achieving this goal has increased.   Factors such as the development of the non-graded school, the various methods of implementing team teaching, the development of self instructional devices, establishment of learning material centers, and various educational experiments, have intensified this effort.   Advantages of individualized instruction, such as saving of student time, child-set learning rate, development of independent study habits, and self-direction, are worthwhile attributes.

A survey of the history of instruction indicates that formal learning began as an individual affair--that is, pupils came to school to receive instruction individually from a teacher.   Circumstances mandated education for a select few; therefore, smaller numbers of pupils attended school. Consequently, individualized instruction was the technique used to teach.   As educational advantages were offered to a larger fraction of the population, individualized instruction diminished.

To achieve sophistication in individualization, development of a curriculum specifying behavioral objectives, the ability to diagnose student needs, and the capacity to prescribe the learning material is mandatory.   This would seem to imply that a program for individualizing instruction necessarily requires the imposition of considerable structure upon the learning situation and on pupil activities.   However, investigations of creativity indicate that self-initiation seems to be encouraged in situations that are not overly detailed in supervision and that do not rely on a prescribed curriculum.   If this is so, and if self-initiation and self-direction are desirable outgrowths of instruction, can self-initiation be an essential element in an individualized program?

Research indicating great differences among individual pupils is reported by researchers who clearly state that just as pupils differ greatly in physical development, motor, intellectual, emotional, and social behavior, they also display wide differences in aptitude and achievement. Fredrick Burk attempted to provide for these differences in aptitude and achievement by pioneering the development of material for individualized instruction.

The Problem

Two classes of elementary school children were carefully studied over a four-month period from January to April, 1966. The study included the use of three instruments to measure self-initiated behavior, student interest, and peer-group evaluation of initiation. Three treatments to encourage self-initiation were introduced during the study. Student behavior was observed by the writer during mathematics, science, and social studies classes. Attempts were made to categorize the source of initiated behavior whether from the teacher, another pupil, or from the student himself. Only the mathematics classes received treatments during the study with the purpose of increasing self-initiation. Student interviews were conducted by the writer at the conclusion of the study to provide insight into student reaction.

Instruments

Three instruments were developed for the study. Instrument One was used by the writer for categorizing the source of initiation for 16 items:

OBSERVERS' RATING SHEET

TEACHER_____ DATE_____
GRADE_____ SUBJECT_____
OBSERVER_____
DIRECTIONS: Place the pupil number from the attached
        seating chart in the proper column for each behavior
        exhibited.
        Self-Initiated: those activities or responses observed
        that do not eminate from teacher direction.

Peer-Initiated: those activities or responses observed that are initiated from another pupil.
Teacher-Initiated: those activities or responses observed that are initiated from the teacher.

	BEHAVIOR OR ACTIVITY	INITIATION ORIGINATING FROM		
		TEACHER	PEER	SELF
(1)	Pupil displays materials brought into classroom. (newspapers, clippings, games, books, collections, etc.)			
(2)	Pupil makes an oral report.			
(3)	Pupil presents a written report.			
(4)	Pupil volunteers to answer questions.			
(5)	Pupil volunteers to work on a committee.			
(6)	Pupil volunteers to do homework or research assignment.			
(7)	Pupil suggests method or solution to problem.			
(8)	Pupil tells about a discussion with parents, friends, or classmates.			
(9)	Pupil tells about a trip or visit.			

BEHAVIOR OR ACTIVITY	INITIATION ORIGINATING FROM		
	TEACHER	PEER	SELF
(10) Pupil tells about T.V. program, movie, etc.			
(11) Pupil goes to pencil sharpener, bulletin board, water fountain, etc.			
(12) Pupil reads book in class.			
(13) Pupil works with supplemental material.			
(14) Pupils request study in special area.			
(15) Pupil presents his material to another class.			
(16) Pupils ask questions.			

Instrument Two was designed to measure student interest in the three school subjects of mathematics, science, and social studies:

PART 1

NAME _____ GRADE _____

SCHOOL _____ DATE _____

DIRECTIONS: Read each statement carefully and circle the answer that tells the way you feel.

(1) When I read for pleasure or information, I read books, newspapers, magazines or stories about: (circle one)
(a) Mathematics   (d) None of the subjects listed
(b) Science   (e) All of the subjects listed
(c) Social Studies

(2) When my teachers ask questions, I volunteer answers in: (circle one)
(a) Mathematics   (d) None of the subjects listed
(b) Science   (e) All of the subjects listed
(c) Social Studies

(3) I like to collect things for: (circle one)
(a) Mathematics   (d) None of the subjects listed
(b) Science   (e) All of the subjects listed
(c) Social Studies

(4) When I do extra work for class, without being told, I do the extra things for: (circle one)
(a) Mathematics   (d) None of the subjects listed
(b) Science   (e) All of the subjects listed
(c) Social Studies

(5) I often talk to my parents about: (circle one)
(a) Mathematics   (d) None of the subjects listed
(b) Science   (e) All of the subjects listed
(c) Social Studies

(6) I talk to my school friends about: (circle one)
(a) Mathematics   (d) None of the subjects listed
(b) Science   (e) All of the subjects listed
(c) Social Studies

(7) If I play school with my friends and family, I like to teach: (circle one)
(a) Mathematics   (d) None of the subjects listed
(b) Science   (e) All of the subjects listed
(c) Social Studies

(8) If I play school with my friends or family, I like to study: (circle one)
(a) Mathematics   (d) None of the subjects listed
(b) Science   (e) All of the subjects listed
(c) Social Studies

(9) I would like to join a club to learn more about: (circle one)
(a) Mathematics   (d) None of the subjects listed
(b) Science   (e) All of the subjects listed
(c) Social Studies

(10) If I could work all day on one subject, I would choose to work on: (circle one)
(a) Mathematics   (d) None of the subjects listed
(b) Science   (e) All of the subjects listed
(c) Social Studies

(11) If I had ten dollars to buy something to help me with my school work, I would spend it for: (circle one)
(a) Mathematics   (d) None of the subjects listed
(b) Science   (e) All of the subjects listed
(c) Social Studies

(12) I would like to watch T. V. programs about: (circle one)
(a) Mathematics   (d) None of the subjects listed
(b) Science   (e) All of the subjects listed
(c) Social Studies

(13) I like to do homework in: (circle one)
(a) Mathematics   (d) None of the subjects listed
(b) Science   (e) All of the subjects listed
(c) Social Studies

(14) If I were the teacher, I would like to teach about: (circle one)
(a) Mathematics   (d) None of the subjects listed
(b) Science   (e) All of the subjects listed
(c) Social Studies

(15) I enjoy writing stories about: (circle one)
(a) Mathematics   (d) None of the subjects listed
(b) Science   (e) All of the subjects listed
(c) Social Studies

Instrument Two

PART 2

NAME   GRADE
SCHOOL   DATE
DIRECTIONS: Read each statement carefully and circle the answer that tells the way you feel.

(1) When I read for pleasure or information. I don't care to read books, newspapers, magazines or stories about: (circle one)
(a) Mathematics   (c) Social Studies
(b) Science   (d) Any of the subjects listed

(2) When my teachers ask me questions, I don't volunteer answers in: (circle one)
(a) Mathematics   (c) Social Studies
(b) Science   (d) Any of the subjects listed

(3) I don't like to collect things for: (circle one)
(a) Mathematics   (c) Social Studies
(b) Science   (d) Any of the subjects listed

(4) I don't do extra work without being told for: (circle one)
(a) Mathematics   (c) Social Studies
(b) Science   (d) Any of the subjects listed

(5) I seldom talk to my parents about: (circle one)
(a) Mathematics   (c) Social Studies
(b) Science   (d) Any of the subjects listed

(6) I don't talk to my school friends about: (circle one)
(a) Mathematics   (c) Social Studies
(b) Science   (d) Any of the subjects listed

(7) If I play school with my friends and family, I don't like to teach: (circle one)
(a) Mathematics   (c) Social Studies
(b) Science   (d) Any of the subjects listed

(8) If I play school with friends or family, I don't like to study: (circle one)
(a) Mathematics   (c) Social Studies
(b) Science   (d) Any of the subjects listed

(9) I would not care to join a club to learn about: (circle one)
(a) Mathematics   (c) Social Studies
(b) Science   (d) Any of the subjects listed

(10) If I could work all day on one subject, I would not choose to work on: (circle one)
(a) Mathematics   (c) Social Studies
(b) Science   (d) Any of the subjects listed

(11) If I had ten dollars to buy something to help me with my school work, I would not spend it for: (circle one)
(a) Mathematics   (c) Social Studies
(b) Science   (d) Any of the subjects listed

(12) I don't care to watch T.V. Programs about: (circle one)
(a)   Mathematics   (c)   Social Studies
(b)   Science   (d)   Any of the subjects listed

(13) I don't care to do homework in: (circle one)
(a)   Mathematics   (c)   Social Studies
(b)   Science   (d)   Any of the subjects listed

(14) If I were the teacher, I would not care to teach about: (circle one)
(a)   Mathematics   (c)   Social Studies
(b)   Science   (d)   Any of the subjects listed

(15) I don't care to write stories about: (circle one)
(a)   Mathematics   (c)   Social Studies
(b)   Science   (d)   Any of the subjects listed

Instrument Three was used to measure peer ratings of the extra work each student contributed to his class:

DIRECTIONS: I would like to know how often you think your classmates do extra things for class. Extra things might be telling the class about places they have visited, or bringing to class newspaper articles or pictures, or it might even be reading a special book about a school subject.

Please put a check mark (✓) across from each name telling me how often you think each classmate does extra things. Do not write your name on this paper.

STUDENTS DO EXTRA THINGS

Name of Student	Never	Almost Never	Sometimes	Often	Very Often

Treatments

The mathematics classes received three general treatments for the purpose of encouraging self-initiation.

Treatment One included the selection of student volunteers from the two classes to help organize a mathematics materials center.   The activity of developing and organizing the mathematics materials center took place outside the regular school day.   Materials gathered and housed in the center were drawn from those available in school and from student contributions.   Student volunteers explained to their classmates what materials were available.   Treatment One was designed to create an awareness of the wide range of supplementary materials available in mathematics and to encourage the use of such materials.

Treatment Two was designed to permit students to explore areas of mathematics that interest them.   Half of the class time scheduled for mathematics was set aside for students to study an area of work other than that assigned by the teacher.   The optional area of mathematics was selected by the students from one of the 12 units of the mathematics continuum in use by the school.   Students were permitted to change their optional areas after the completion of a unit of work.

The purpose of Treatment Three was to provide special reinforcement or rewards for the students during the mathematics classes.   Each teacher made a concentrated effort to praise exceptional work and to display student materials.   Seminar classes provided an opportunity for students to review their mathematical interests with other class members.   Students used this opportunity to explain their special interest areas, display their work, and review findings.

Treatment One was introduced during the first month of the study and was in effect during the entire study.   Treatment Two was introduced during the second month of the study and again was in effect during the remaining portion of the study.   Treatment Three was introduced during the third month of the study.   It can be noted that all treatments, once begun, continued throughout the research.

The study was conducted over a four-month period, January to April, during the school year 1965-66.   Each of the two classes in mathematics, science, and social studies were observed eight times during this period when Instrument One was used.   The time between the series of observations was devoted to the application of the three treatments.

Classroom Observations

        The fifth- and sixth-grade classes of mathematics,
science, and social studies were observed eight times during
the study using Instrument One. Two observations were conducted for each class during the first, sixth, 11th, and 16th
weeks of the study. The student population of the fifth-grade
class was 27 and the sixth grade, 22, thus making a total of
49 students observed.

        The purpose of the classroom observations was to
credit each student for behavior exhibited as defined by Instrument One. Furthermore, the source of stimulation for
the exhibited behavior was categorized. Each time a student
exhibited any of the behaviors or activities listed in Instrument One, one point was credited to that student. If, for
example, Student A was observed volunteering to do a homework or a research assignment, he was given one credit for
that behavior. Furthermore, if the stimulation for doing
homework appeared to be related to the teachers, it was
categorized in the teacher-initiated column. If the student
volunteered to do a homework or research assignment because another pupil suggested that this information was necessary, the behavior was categorized as peer-initiated. If,
however, the student volunteered to do homework or a research assignment without teacher or peer stimulation, it
was categorized as self-initiated. When the teacher assigned
work to the whole class, each member was credited for this
activity under the proper category. Therefore, the numbers
that appear in the tables that follow represent the count taken
for each item appearing in Instrument One, and are categorized in the proper column of teacher-initiated, peer-initiated,
or self-initiated.

        The first item listed in Instrument One was "pupils
display materials, " such as newspapers, clippings, games,
books, or collections brought into the classroom. This behavior was observed to take place 16 times--eight incidents
in each of the social studies and science classes. Pupils
display of materials did not occur at all during the mathematics classes. Three of the 16 recorded observations referred to above were directly related to teacher-initiation,
while 13 of the behaviors were credited as self-initiated.
The data for this first item appears in Table 1.

TABLE 1

Data Obtained From Instrument One:
Item One--Pupils Display Materials

Class	Teacher-Initiated	Peer-Initiated	Self-Initiated	Total
Social Studies	0	0	8	8
Science	3	0	5	8
Mathematics	0	0	0	0
TOTAL	3	0	13	16

Item two listed in Instrument One referred to oral reports by pupils. This behavior was observed to take place 55 times during the study--49 incidents that were categorized as self-initiated and six as teacher-initiated. Thirty-eight incidents occurred in the social studies classes and 17 in the science classes. No incidents of pupils making oral reports were observed in the mathematics classes.

TABLE 2

Data Obtained From Instrument One:
Item Two--Pupils Make An Oral Report

Class	Teacher-Initiated	Peer-Initiated	Self-Initiated	Total
Social Studies	0	0	38	38
Science	6	0	11	17
Mathematics	0	0	0	0
TOTAL	6	0	49	55

Item three in Instrument One was "pupils present a written report." This behavior was observed 19 times during the study--ten in the social studies classes, and nine in the science classes. No incidents of pupils presenting written reports were observed in the mathematics classes. Eleven behaviors were credited as teacher-initiated, two as peer-initiated, and six as self-initiated.

Item four in Instrument One was "pupils answer questions." This behavior was observed 2,052 times during the study. Sixty-one incidents of the observed behavior were categorized as self-initiated, and 126, peer initiated. Pupils answering questions tended to be teacher-initiated with 1,865 recorded incidents in this category. Table 3 indicates that 813 incidents were recorded for social studies, 793 in the science classes, and 446 in the mathematics classes.

TABLE 3

Data Obtained From Instrument One:
Item Four--Pupils Answer Questions

Class	Teacher-Initiated	Peer-Initiated	Self-Initiated	Total
Social Studies	713	66	34	813
Science	709	58	26	793
Mathematics	443	2	1	446
TOTAL	1,865	126	61	2,052

Item five in Instrument One was "pupils volunteer to work on a committee." This behavior was observed 64 times during the study, occurring 21 times in the science classes and 43 times in the social studies classes. Teacher-initiation was credited with 42 incidents and self-initiation, 20. Pupils volunteering to work on a committee were not observed during visits to the mathematics classes.

Item six in Instrument One was "pupils volunteer to do homework or research assignments." This behavior was observed 540 times during the study. Forty-one incidents were categorized as self-initiated, while 497 were teacher-initiated. No incidents of this behavior were credited to mathematics, 248 incidents were credited to the social studies classes and 292 to the science classes.

Item seven in Instrument One was "pupils suggest methods or solutions to problems." This behavior was observed 34 times during the study, occurring 19 times in social studies and 14 in the science classes. This behavior was only observed once in mathematics. During the individualized mathematics classes, it was almost impossible to know when one student, talking privately with the teacher,

was suggesting methods or solutions to problems.   Generally
solutions suggested by the students in the science and social
studies classes were, "We can find the answers in the encyclopedia. "   Thirty-two of the observed behaviors were
credited as self-initiated, and two were credited as teacher-
initiated.

Item 11 in Instrument One was "pupils read a book in
class. "   This behavior was observed 344 times during the
study, 167 incidents being credited to social studies, 153 to
science, and 24 to mathematics.   When the teacher directed
the class to open their books to a certain section, each student was credited for this behavior and it was categorized as
teacher-initiated.   As one would suspect, this activity occurred most often as teacher-initiated.   The mathematics
and science classes produced the highest incidents of self-
initiation for item 11.

TABLE 4

Data Obtained From Instrument One:
Item 11--Pupils Read A Book In Class

Class	Teacher-Initiated	Peer-Initiated	Self-Initiated	Total
Social Studies	162	0	5	167
Science	126	0	27	153
Mathematics	1	0	23	24
TOTAL	289	0	55	344

Item 12 in Instrument One was "pupils work with supplemental materials. "   Supplemental materials were those
learning tools used by the students that are not usually part
of the lesson materials.   For example, maps, globes, film
strips and counting frames were considered as supplemental
materials.   However, the regular textbooks and work pages
were not considered supplemental.   This behavior was observed 303 times during the study.   The social studies
classes produced a total of 199 incidents of this behavior,
of these 183 were teacher-initiated.   The mathematics
classes produced 88 incidents of this behavior, all being self-
initiated.   The science classes produced the fewest number
of incidents of pupils working with concrete materials.

TABLE 5

Data Obtained From Instrument One:
Item 12--Pupils Work With Supplemental Materials

Class	Teacher-Initiated	Peer-Initiated	Self-Initiated	Total
Social Studies	183	0	16	199
Science	4	0	12	16
Mathematics	0	0	88	88
TOTAL	187	0	116	303

Item 13 in Instrument One was "pupils request study in a special area." This behavior was observed six times during the study, occurring four times in the mathematics classes and twice in social studies. As might be expected, all of the observed behavior for this item was categorized as self-initiated.

Item 14 in Instrument One was "pupils present their materials to another class." This behavior never occurred during the observation.

Item 15 in Instrument One was "pupils go to the library, clerks, or scoring keys." This behavior was observed 1,465 times during the observations. During the mathematics classes 1,410 incidents of students going to the clerks with materials to be corrected or obtaining self-scoring materials was categorized. Although this behavior might be considered part of the procedure of the individualized mathematics program, students did initiate such behavior without teacher direction. Therefore, the 1,406 times when this behavior occurred during the mathematics classes, without specific teacher direction, it was credited as self-initiated. This accounts for the larger self-initiation in mathematics when compared with the science and social studies classes.

Item 16 in Instrument One was "pupils ask questions." This item was not included as part of Instrument One during the first week of observations. Therefore, the results are based on three weeks of observations, or a total of six class visits per subject. The behavior was observed 977

TABLE 6

Data Obtained From Instrument One:
Item 13--Pupils Request Study In A Special Area

Class	Teacher-Initiated	Peer-Initiated	Self-Initiated	Total
Social Studies	0	0	2	2
Science	0	0	0	0
Mathematics	0	0	4	4
TOTAL	0	0	6	6

TABLE 7

Data Obtained From Instrument One:
Item 15--Pupils Go To Library, Clerks, Scoring Keys

Class	Teacher-Initiated	Peer-Initiated	Self-Initiated	Total
Social Studies	22	3	25	50
Science	2	0	3	5
Mathematics	4	0	1,406	1,410
TOTAL	28	3	1,434	1,465

times, 928 incidents being credited as self-initiated.   Students exhibited seven times as many incidents of asking questions in mathematics than they did in the science or social studies classes.   As indicated in Table 8, this behavior was observed 118 times in the social studies classes, 137 in science, and 722 in mathematics.   Questions asked by students that were in direct response to the teacher saying, "Are there any questions?" were categorized as teacher-initiated.   Based on the six observations per subject and the 49 students observed, the average number of questions observed being asked was 19.6 in social studies, 22.8 in science, and 120.3 in mathematics.

Table 9 displays a summary of the 16 items listed in Instrument One for the three subjects of social studies,

TABLE 8

Data Obtained From Instrument One:
Item 16--Pupils Ask Questions

Class	Teacher-Initiated	Peer-Initiated	Self-Initiated	Total
Social Studies	10	7	101	118
Science	29	3	105	137
Mathematics	0	0	722	722
TOTAL	39	10	928	977

science, and mathematics.

The 16 items that were listed in Instrument One represent several aspects of student behavior that are observable. With the exception of item 14 relating to pupils presenting material to another class, behavior was observed for the other 15 items. This was particularly true of the science and social studies classes. Fifteen items were recorded for the social studies classes and 13 for the science classes. Seven of the items were recorded for the mathematics classes. Either these behaviors were not observed or were impossible to record due to the nature of the individualized mathematics program. For example, it was not always possible to hear the conversation between the teacher and the individual student; therefore, such items as suggesting solutions to problems, telling about a trip, or discussions with parents were not observed during the mathematics classes. These items were more easily observed in classes that were taught in group situations.

Combining all 16 items from Instrument One, for all observations, indicated which classes were more teacher-initiated, peer-initiated, or self-initiated. Appearing in Table 10 are the total observation scores for the three classes; it also shows the percentage of initiated behavior for each category.

It should be noted that item 15 listed in Instrument One categorized 1,406 behaviors as self-initiated for the individualized subject of mathematics. Therefore, of the total 2,245 behaviors categorized as self-initiated in mathematics,

TABLE 9

Summary Of Data Obtained From Instrument One For
Social Studies, Science And Mathematics

ITEMS	TEACHER-INITIATED			PEER-INITIATED			SELF-INITIATED			TOTAL		
	SS	SC	M	SS	SC	M	SS	SC	M	SS	SC	M
1	0	3	0	0	0	0	8	5	0	8	8	0
2	0	6	0	0	0	0	38	11	0	38	17	0
3	7	4	0	1	1	0	2	4	0	10	9	0
4	713	709	443	66	58	2	34	26	1	813	793	446
5	21	21	0	2	0	0	20	0	0	43	21	0
6	207	290	0	2	0	0	39	2	0	248	292	0
7	2	0	0	1	0	0	17	14	1	29	14	1
8	0	1	0	0	0	0	10	2	0	11	3	0
9	3	1	0	0	1	0	19	10	0	22	11	0
10	0	1	0	0	0	0	1	8	0	1	10	0
11	162	126	1	0	0	0	5	27	23	167	153	24
12	183	4	0	0	0	0	16	12	88	199	16	88
13	0	0	0	0	0	0	2	0	4	2	0	4
14	0	0	0	0	0	0	0	0	0	0	0	0
15	22	2	4	3	0	0	25	3	1406	50	5	1410
16	10	29	0	7	3	0	101	105	722	118	137	722
TOTAL	1330	1197	448	82	63	2	337	229	2245	1749	1489	2695

TABLE 10

Data Obtained From Instrument One:
Total Of All Items and Percentages

CLASS	TEACHER-INITIATED		PEER-INITIATED	
	NO.	%	NO.	%
Social Studies	1, 300	76. 04	82	4. 69
Science	1, 197	80. 39	63	4. 23
Mathematics	448	16. 63	2	. 07

CLASS	SELF-INITIATED		TOTAL	
	NO.	%	NO.	%
Social Studies	337	19. 27	1, 749	100
Science	229	15. 38	1, 489	100
Mathematics	2, 245	83. 30	1, 695	100

1, 406 are from one item. It could be argued that this behavior is part of the procedures for the individualized classes
and should be considered as teacher-initiated. If this were
done, the percentage for teacher-initiated and self-initiated
behavior listed in Table 10 would change. The mathematic
classes would be approximately 68% teacher-initiated and
31% self-initiated. This serves to indicate that the individualized classes would still be more self-initiated than the
non-individualized classes.

The first hypothesis postulated that there would be an
increase in self-initiated activities in the individualized mathematics classes following the introduction of the treatments.
Appearing in Table 11 are the data from Instrument One for
the self-initiated scores for the mathematics classes. The
data are organized by observations for each week. Notice
that seven items were observed for the mathematics classes
and that four items account for most of the activity.

The first treatment which was the development of a
mathematics materials center was introduced after the first

TABLE 11

Self-Initiated Scores Derived From The Use Of
Instrument One in Mathematics Classes

Item	OBSERVATION WEEKS				
	Week 1	Week 2	Week 3	Week 4	Total
1	0	0	0	0	0
2	0	0	0	0	0
3	0	0	0	0	0
4	0	1	0	0	1
5	0	0	0	0	0
6	0	0	0	0	0
7	0	0	0	1	1
8	0	0	0	0	0
9	0	0	0	0	0
10	0	0	0	0	0
11	0	9	7	7	23
12	0	35	31	22	88
13	0	0	4	0	4
14	0	0	0	0	0
15	396	437	326	247	1,406
16	---	285	265	172	722
TOTAL	396	767	633	449	2,245
TOTAL less item 16	396	482	368	277	1,523

week of observation and continued in effect during the course
of the study.   Item 11, which was "pupils read a book in
class," increased nine incidents between the first and second
weeks of observations.   This behavior continued during the
third and fourth weeks of observations with seven incidents
being credited for each of the two weeks.   Supplementary
reading material related to mathematics was part of the
mathematics materials center.   Item 12, "pupils work with
supplemental materials," increased 35 incidents after the introduction of Treatment One.   This behavior continued during
the third and fourth weeks of observation, but at a reduced
rate.   That is, the behavior was observed 31 times during
the fourth week of observation.   Item 16 was not recorded

during the first week of observation and, therefore, cannot
be included as evidence of increased self-initiation.

The Second Treatment permitted students to select an
optional area of mathematics.   Half of the time scheduled
for mathematics was devoted to optional work.   Treatment
Two occurred between the second and third weeks' observations and continued in effect during the entire study.

As indicated in Table 11, the number of units of work
completed by the students for the optional areas was 69,
while 128 units of work were completed for the prescribed
areas.

Treatment Three, which was applied between the third
and fourth series of observations, had teachers make a concentrated effort to praise exceptional work during mathematics classes and display student material.   No evidence of
the success of this treatment was observable through the use
of Instrument One.

The increase in the use of concrete materials and the
use of supplementary books suggests that Treatment One affected the self-initiated activities during the mathematics
classes.   It further suggests that, if the use of supplemental
materials is an important behavior, specific provisions must
be made within the classroom to encourage this behavior.

More units of work were completed when teachers
prescribed the mathematics units.   Although 69 optional units
were completed during the study, 128 units were completed
by teacher assignment.   The skills within a unit of work are
not equal in number or difficulty.   Therefore, it is not
known if the 69 optional units represents less work on the
part of the students than the 128 units of prescribed work.
Also, it is not known if a total of 197 units of work would
have been completed by the students if optional work had not
been introduced.   It was the writer's observations that the
students worked harder and asked more questions during optional days.   Also, students tended to select more difficult
units of work during optional days.   Geometry and the study
of other member bases were very popular.   With the exception of students requesting work in special areas, Instrument
One did not reveal any other effect of Treatment Two.

The increase of self-initiated behavior of items 11 and
12, and the additional requests for study in a special area

suggests that there was an increase in specific self-initiated activities which the treatments were designed to enhance in the individualized mathematics classes. Therefore, hypothesis one is not rejected.

The second hypothesis stated, "There will be no noticeable increase in self-initiation in the untreated subject of social studies."

During the first week of observations a total of 57 self-initiated behaviors were observed. The second week of observations produced a total of 156 self-initiated behaviors. This included 47 behaviors for item 16 which was not included during the first week of observations. The total self-initiated score for the third week of observation was 80, including 35 incidents for item 16. The last observation produced a total of 46 self-initiated activities, including 19 for item 16. Based on the score of 57 for the first week of observations and a total score of 27 (not including item 16) for the last week of observations, it is apparent that there was no increase in self-initiation in the social studies classes. Therefore, hypothesis two is not rejected.

Hypothesis three stated, "There will be no noticeable increase in self-initiation in the untreated subject of science." The total self-initiated score for the first week of observations was seven. This total increased to 37 during the second week, 59 during the third week, and 19 for the fourth week. Item 16 is not included in the totals since it was not included for the first week of observations. These data suggest that self-initiation did increase during the study. Therefore, hypothesis three is rejected. Note that item 11, "pupils read a book in class," increased from the first to the last observation and was substantially higher during the third week. This occurred when the teacher introduced a library of science material to the class for independent work.

Hypothesis four stated, "There is a significant relationship between IQ score and self-initiation before and after treatments are applied." Appearing in Table 12 is the correlation of the self-initiation score for mathematics, before and after the treatments. It can be seen that there is no substantial correlation between IQ score and self-initiation. Therefore, this hypothesis is rejected.

Hypothesis five indicated, "There is a significant relationship between scores on standardized achievement tests

TABLE 12

Correlations Of IQ With Self-Initiation Scores For
Mathematics Before and After Treatments*

Observation	N	Correlation of IQ With S-I Score In Mathematics
Before Treatment	49	.15
After Treatment	49	.21

*Neither of the correlation coefficients presented differ significantly from zero.

in mathematics and self-initiation. " In Table 13 are the data derived from correlation of standardized achievement test scores for mathematics concepts and problem solving with the final self-initiation score. The correlation score suggests substantially no correlation. Therefore, this hypothesis is rejected.

TABLE 13

Correlations Of Self-Initiation Score With Standardized
Mathematics Score For Problem Solving And Concepts*

Area	N	Correlation
Problem Solving	49	.165
Concepts	49	-.002

*Neither of the correlation coefficients presented differ significantly from zero.

Hypothesis six stated, "There is no significant relationship between sex of students and self-initiation. " Using the final self-initiation score and correlating it with the sex of students gives a correlation of .124. Therefore, this hypothesis is not rejected.

Hypothesis nine stated, "There is a significant relationship between teacher rating and peer-evaluation of self-
initiation. " The correlation score between the teacher rating
and the peer-evaluation was . 56 before treatment and . 70
after treatment.   Therefore, this hypothesis is not rejected.

The three treatments introduced into the study as attempts to increase self-initiation had limited success.   The
first treatment, development of a mathematics material center, seemed to create more self-initiation in the mathematics
classes.   Students did use many of the materials as was
evidenced during the classroom observations and the increase
in several of the items listed on Instrument One.   The second treatment, permitting students to select optional units in
mathematics seemed to encourage self-initiation, although
documented data for this treatment was more difficult to obtain.   However, the selection of more difficult work in
mathematics and the increased number of questions students
ask during optional days suggest the treatments did provide
for more self-initiation.   The third treatment, providing reinforcement to students during mathematics classes, appears
to have been less successful.   During the student interviews,
it was revealed that this treatment reached about 50 per cent
of the fifth-grade class and 20 per cent of the sixth-grade
class.   It must be noted that the procedures used in the
mathematics classes increase student-teacher contact and
provide for daily reinforcement.   Therefore, it can be concluded that the first treatment was most effective and the
last treatment the least.

The interviews conducted revealed that students believed that the procedures used in teaching mathematics provided an opportunity for students to work on their own, to
teach themselves, and to go at their own speed.   These
aspects of the procedures used to teach the individualized
mathematics were seen as favorable by the pupils.   The
social studies classes were best liked because of the opportunity to do project work and the nature of the subject itself.
When students were asked to choose between an individualized
science program, which they had had previously, and a non-
individualized class, the older students selected the individualized approach.   The fifth-grade students were divided as
to their selection.

The students stated that the mathematics center was
helpful and expressed a wish to have it continued.   Optional
work in mathematics was felt to be worthwhile, and students

expressed a desire to continue the optional program.

From the analysis of the nine hypotheses and the student interviews, we may conclude that:

(1) Individualized instruction seems to be more self-initiated than non-individualized.

(2) The amount of self-initiation in a classroom can be increased by the introduction of specific techniques to improve this activity.

(3) Self-initiation has little relationship to intelligence, achievement, or sex of students.

(4) Expressed interest in the subject of mathematics did not change over the four months of the study.

(5) The treatments had no measurable effect on expressed interest.

(6) The procedures used to encourage self-initiation in the individualized classes had little carry over to the non-individualized classes.

(7) The teacher ratings of the amount of extra activity students do for school had a correlation range between moderate to high with the student ratings of each other.

(8) The student ratings of extra school activity of each other did not change over the four-month period.

(9) The pupils expressed a desire to continue with some of the treatments in their mathematics classes.

(10) The students hoped to obtain a professional occupation with "teacher" ranking high.

MEASURING LEARNING RATE IN AN IPI PROGRAM

John L. Yeager

The purpose of this study is to investigate the consistency over different units of study of three measures of rate of classroom learning in elementary school mathematics and reading, and to ascertain the relationship of these measures to selected student variables.

In order to reduce the degree of ambiguity that can exist in discussing the variables that are employed in this study the following definitions are stated.

RATE OF LEARNING: three definitions of rate are used in this study: the number of units of work that a student completes within a period of one year; the number of days a student requires to complete a given unit of work; and the average daily achievement of a student working on a particular unit of work as represented by the equation:

$$\text{Rate} = \frac{100 \text{ per cent minus achievement on pre-test}}{\text{Number of days in unit}}$$

IPI (Individually Prescribed Instruction): a specific program for the individualization of instruction in selected elementary school subjects, developed as a project of the Learning Research and Development Center in collaboration with the Baldwin-Whitehall schools and the staff of the Oakleaf Elementary School.

CONSISTENCY: three measures of consistency are used in this study: the Pearson product-moment correlation between and within curriculum areas; the F-test for determining the significance of the differences among individual students when rate measures are totaled over several units; and Hoyt's estimate of the reliability of a total measure using an analysis of variance approach.

READING ACHIEVEMENT: grade equivalent scores obtained on the Metropolitan Achievement Test Battery in the area of reading comprehension. Grade equivalent scores were used instead of raw scores because of the different forms used for the various grade levels.

Review of Related Research

A concern that has been increasingly evident in educational theory and practice during recent years is for the development of instructional programs that make some provision for individual differences. One evidence of this concern is the increasing volume of professional literature pertaining to problems that are confronted in any program for individualized instruction. It is also evidenced in the development of a variety of plans for individualization such as those utilizing programmed instruction, ability grouping, independent study, team teaching, enrichment and remedial programs, and non-graded instructional plans.

This heightened activity in developing new methodologies and organizational plans is partly a result of an

increased awareness of the wide range of inter- and intra-
individual differences found in any classroom. Goodlad and
Anderson, [1] in a comprehensive study of non-grading, summarize six generalizations concerning the student and the
conventional graded system of instruction:

(1) Children enter the first grade with a range of
from three to four years in their readiness to profit from a
"graded minimum essentials" concept of schooling.

(2) This initial spread in abilities increases over the
years so that it is approximately double this amount by the
time children approach the end of the elementary school.

(3) The achievement range among pupils begins to
approximate the range in intellectual readiness to learn soon
after first-grade children are exposed to reasonably normal
school instruction.

(4) Differing abilities, interests, and opportunities
among children cause the range in certain specific attainments to surpass the range in general achievement.

(5) Individual children's achievement patterns differ
markedly from learning area to learning area.

(6) By the time children reach the intermediate elementary grades, the range in intellectual readiness to learn
and in most areas of achievement is as great as or greater
than the number designating the grade level.

Cook and Clymer, [2] supporting these generalizations in
their findings concerning elementary school children, state
that the range of achievement in given grades follows rather
regular patterns. If the 2 per cent at each extreme of the
distribution is eliminated, the range of ability represented
by a class is two-thirds of the chronological age of the typical student in the grade.

These findings indicate the wide variability of student
achievement that is typically found in the conventional-type
classroom.

It also would be anticipated that variability would be
evidenced in terms of student rate of progress. This particular aspect of the learning situation has to date received
only limited attention. The variable, "rate of learning, " has
traditionally been studied in the laboratory since typical classroom procedures do not yield information concerning rate of
student progress. Traditional methodologies have attempted
to disregard individual rates of learning by structuring the
learning situation into a series of common experiences that

attempt to have all students progress at a common rate.   It
has only been since the recent advent of programmed instruction, the non-graded organizational plan, and individualized
instructional programs that individualization in rates of progress has been facilitated.   Many investigators have shown
that students, when given the proper opportunity, do progress at varying rates.

Suppes, [3] working with 40 first-grade students, with
a mean intelligence quotient of 137, under an essentially individualized method of instruction, found a wide divergence
in student learning in mathematics.   Suppes states that the
most significant aspect of this individualized treatment was
the fantastic difference in rate of learning.   At the end of
the first four weeks of this program, the fastest student had
covered 150 per cent more material than was covered by the
slowest student.

Kalin, [4] in developing a programmed text in mathematics for superior fifth- and sixth-grade students, tested his
mathematics program on an experimental and control group.
His findings indicated that the experimental group consisting
of 95 students achieved the same degree of mastery as the
control group but completed the work in 20 per cent less
time.

These studies indicate that when the student is provided the opportunity to progress at a self-determined rate
a wide variance in learning rates is found.   It is of importance to note that this variance was evidenced under different
methods of instruction and does not necessarily appear to be
unique to a particular type of instruction.

There have been relatively few studies undertaken to
determine the characteristics of various measures of rate of
learning, although a great deal of theorizing has been done.

One early study that tends to support these hypotheses
was undertaken by Woodrow[5] who spent 20 years studying the
concept of learning rate in the laboratory through the analysis
of learning curves.   These studies were carried out under
controlled conditions and used the slope of the learning curve
as an estimate of learning rate.   His findings indicate that
there exists no general factor acting as one of the determining conditions of rate of improvement in widely different
types of performance.   This conclusion holds whether rate
of improvement is measured by absolute gain, the gain per

unit of time at the point on the curve when the individual's
gain is at its maximum, or relative gain, the same gain
taken as a proportion of the individual's indicated ultimate
score.

Carroll[6] has proposed a learning model on the assumption that the individual will succeed in learning a given
task to the extent that he spends the amount of time needed
to learn the task. In this instance, the learning task is defined as "going from ignorance of some specific fact or concept to knowledge or understanding of it, or proceeding from
incapability of performing some specific act to capability of
performing it. " Time, as measured in this model, is not
elapsed time but rather time actually spent in paying attention and trying to learn. There are certain factors, such
as aptitude and quality of instruction, which determine how
much time an individual needs to spend in order to learn
the task. These factors may or may not be the same as
or associated with those influencing how much time he spends
in learning. Carroll summarizes his model by the equation:

$$\text{Degree of Learning} = F \frac{(\text{amount of time actually used})}{(\text{time needed})}$$

The main problem in utilizing Carroll's model is the difficulty in quantifying the necessary variables.

Most of the studies that have been undertaken use
Ebel's[7] definition of time in obtaining measures of learning
rate: "Rate is the measure of an individual's speed of performance on tasks of a particular type, stated either in
terms of the number of units of work done in a given time
or the number of units of time required to complete a given
amount of work. "

Unfortunately, there has been little work done to determine the reliability of these various measures of rate.
The one study that considers the problem of reliability of a
measure of rate was carried out by Gropper and Kress[8] in
studying the effects of different patterns of pacing student
work using programmed materials. In this study, two groups
of eighth graders were administered two science programs
of approximately 100 frames in length. One of the most
striking findings of this study was the consistency of student
pace or work rate from program to program. Rate, in this
instance, was the amount of time needed to complete the
program. A correlation of . 80 was obtained between the

rate measure for the two programs.

While this specific study demonstrated a reliable measure of rate for the particular type of instruction and task involved, other measures have not yielded such reliable results.   It should, therefore, be important to obtain reliability estimates pertaining to other measures of rate under different methods of instruction for a variety of specified tasks.

The determination of specific student characteristics that influence different measures of learning rates if of particular interest.   One characteristic, that of the individual's intelligence quotient, has received a relatively large amount of attention.   Woodrow, [9] in one of the most comprehensive studies undertaken, obtained results that indicated no relationship between rate of improvement and intelligence.

In an investigation of learning ability by Jensen, [10] junior high school students classified as "educationally mentally retarded" and having Stanford Binet IQ scores from 50 to 75 were compared on a selective learning task with average IQ (scores from 90 to 110) and gifted IQ (scores above 135) children in the same school.   The task consisted of learning by trial-and-error to associate five or six different stimuli, colored geometric forms, with five or six different responses.   The responses in this study consisted of an array of push buttons.   Jensen developed an index of learning that was used to indicate the student's rate of learning. This index can be interpreted as the percentage of maximal possible performance above the level of chance performance. His findings indicated that there were highly significant differences among the groups and that the student's rate of learning correlated with intelligence.   This finding was in agreement with the findings of the Glaser, Reynolds, and Fullick[11] investigation that studied intensively the effects of programmed instruction under a variety of conditions with students in different grades.   Their findings indicate, in general, that intelligence appears to be related to the pace with which the student works through the program.

These findings of Jensen and the Glaser, Reynolds and Fullick study are in contradiction with those of Woodrow and others.   This discrepancy probably is the combined result of the interactions of the types of rate measures employed, tasks that were learned, and the instructional methods that were utilized.

The purpose of this study will be to investigate some of the qualities of various measures of rate of classroom learning. Obviously, this type of study can be pursued only in a classroom situation which permits pupils to proceed through a learning sequence at individual rates. This provision is an essential part of the program for IPI--the instructional program in which this study was carried out.

This program of individualized instruction will permit the investigation of various types of rate measures such as days in unit, number of units completed in a given time period, and various indexes of learning rate. These rate measures can then be examined as to their consistency between units in the mathematics and reading sequences and their relationship to selected student characteristics. Also, attention can be focused on the relationship between rates in different tasks within and between the curriculum.

Measures of Rate of Learning

There are a variety of rate measures that could have been investigated, but this study was concerned with only three rather basic and simple measures. The first measure was the total number of mathematic and reading units mastered by the student during the school year. Although this was a rather obvious measure of learning rate, it does provide information concerning the student's progress over an extended period of time. The primary limitation in using this type of measure was that the units being studied were not of equal complexity or length.

In order to partially compensate for this problem of unequal complexity or length of the units, a second measure was studied consisting of the number of days a student required to master a given unit. This measure represented a more refined measure of learning rate in that it considered only the amount of time a student spent on a given unit. A major limitation of this measure was that it did not take into account the student's knowledge of the material before starting to work in the unit.

Because of this limitation, a third measure was studied that consisted of an index of rate of learning. To obtain an index of student rate of learning for a particular unit of work the pre-test score was subtracted from 100 and divided by the number of days spent in the unit. The index,

therefore, represented the student's average daily achievement for a given unit of work.   This measure considered
the amount of content a student mastered in a unit and partially controlled for the student's entering behavior.   One
hundred was selected as the criterion because when a student
was assigned a particular prescription, this prescription
theoretically contained a sufficient number of learning experiences for the student to obtain 100 per cent mastery.

Research Population

The research population that was used consisted of
students in grades one through six at the Oakleaf Elementary
School.   A breakdown by grade of the number of students
participating in this study is shown in Table 1.

TABLE 1

Number Of Students Per Grade

Grade	1	2	3	4	5	6	Total
Number of Students	27	26	30	26	21	23	152

Statistical Procedures

The statistical analyses contained in this study were
such as (1) to provide information concerning the reliability
or consistency of each of the three measures of learning
rate, and (2) to provide information concerning the relationship of these measures to such student characteristics as intelligence and reading ability.   To obtain this information
and permit a comparison of the three measures on these
qualities, the following analyses were made:

(1)   The correlation between rate measures for various subjects or sub-areas within a subject were computed in
connection with each measure.   This provided an index of
the consistency of the measure over different subjects or
units.

(2)   Where possible, namely with the second and third

measures of rate, consistency was further studied through the use of analysis of variance.   This involved testing the significance of differences among students in rate measures obtained over several units of study.   This analysis also provided for the determination of the reliability of total rate measures obtained over several units through the use of the procedures suggested by Hoyt.

(3)   Each of the three rate measures were further studied by determining the correlation between such measures obtained over a number of units of instruction and the student characteristics of intelligence and reading ability.

Summary

Number of Units Completed in One Year

The first measure of student learning rate studied was the number of mathematics and reading units a student mastered during one year of work.   Of initial interest was the consistency of this measure in terms of the correlation between the number of mathematics and the number of reading units a student completed during one year of study.   A correlation coefficient of $+.31$ was obtained that was significant at the $.01$ level.   However, this does not indicate that there is a large proportion of shared variance between the two variables.

The relationship between this measure, number of units completed in the areas of mathematics and reading and student intelligence were examined.   The data indicate a slight but significant relationship between the number of mathematics units completed and student intelligence.   This finding, however, was not duplicated when the number of reading units was correlated with student intelligence.

As an extension of the study of the relationship between student intelligence and the number of units mastered, the relationship between student intelligence and level of initial and final placement in the IPI sequence was investigated.   Since rate has been measured in terms of number of units, level of initial and final placement were measured in the same manner.   The correlation coefficient obtained indicates a moderate relationship between intelligence and the amount of content that a student has mastered in mathematics and reading prior to entering the IPI program.   A moderate

to strong relationship was also evidenced between student intelligence and the level of final attainment in the mathematics
and reading curriculums.

These findings indicate that there does exist a moderate relationship between student intelligence and rate of
learning when this is measured in terms of units of work.
It is of interest to note that there is a moderate relationship
between student intelligence and student initial and final
placement in the mathematic and reading sequences and only
a slight relationship between intelligence and total number of
units completed.

The second student characteristic studied in conjunction with this measure of learning rate was level of student
reading achievement.   When this student variable was correlated with the number of mathematics united completed during
one year a slight but significant relationship was evident.
This relationship was not found when the reading curriculum
was analyzed.   One possible reason for these conflicting
findings could be a function of the more restricted range of
units mastered in the reading curriculum.

This measure of learning rate, number of units mastered, does seem to be related to the student characteristics
of intelligence and level of reading achievement in the case
of rate in mathematics but not in reading.

In considering number of units completed in a school
year as a measure of rate of learning certain points should
be noted.   First, of course, that it is a practical and meaningful measure in that it indicates how quickly a pupil can
progress through a sequence of learning experiences when
all the factors in the classroom that can effect student progress are allowed to operate.   A second point is that, in using the number of units completed as a measure of learning
rate, it is essentially impossible to make the units equivalent
in difficulty or in the average time needed for mastery.
Therefore, because of these complicating factors, additional
measures were studied.

Number of Days to Master a Given Unit

To avoid the problem of varying unit difficulty a
measure of student rate of learning in terms of number of
days to complete a given unit was investigated.   The

consistency of this rate measure was investigated between different topics at the same level, the same topic at different levels, over three or more topics at varying levels, and between the mathematics and reading curriculum.

The first situation, that of the consistency of the measure, number of days per unit between topics at the same level, resulted in few significant correlation coefficients in either mathematics or reading. In general there was no consistency between the rate in one topic and another topic at the same level.

In studying the consistency between the same topics at different levels, a smaller number of samples were used because of the limited number of students completing enough units to cover more than one level of work. The correlation coefficients resulting from this analysis indicated a lack of consistency between the same topics at successive levels of complexity in both the mathematics and reading curriculum. However, caution should be exercised in drawing any definite conclusions because of the limited number of units involved.

The final study of consistency involved a correlational study between units of the mathematic sequence and the reading sequence. When rate measures in units of mathematics were correlated with those in reading there were, in general, no significant correlation coefficients. This appears to be in conflict with the finding that there was a significant relationship between the total number of mathematics and reading units completed in a period of one year.

This can be explained in terms of the type of rate measures that were employed. When the rate measure of days to complete unit was used, a single unit of the mathematics sequence was compared to a single unit of reading. In this manner the relationship between two specific tasks, one in mathematics and one in reading, was studied. When student rate was investigated in terms of the number of units completed in mathematics and reading, no consideration was given to the wide variety of tasks represented by this measure. In addition to this, the units were nonequivalent in difficulty or length. While the two measures are both measures of rate of learning, they are measuring two somewhat different things.

The second student variable that was investigated in conjunction with the rate measure number of days to complete unit was that of level of student reading achievement. The findings indicate that there is no general relationship between the time required to complete a given unit of mathematics or reading and student level of reading achievement. This result is interesting because, for the most part, the student learning experiences that were prescribed consisted of materials that required a great deal of reading.

Average Daily Achievement

The third measure of rate of learning to be studied was that of average daily achievement. This measure takes into account the entering behavior of the student by dividing the amount a student achieved in a unit by the time required to complete the unit. This rate measure was analyzed in the same manner as the previous measure, number of days to complete a unit.

The consistency of this measure of learning rate was studied in terms of the consistency between different topics at the same level, consistency between the same topics at different levels, consistency over three units, and consistency between the curriculum areas of mathematics and reading.

In investigating the first of these four relationships, consistency between the different topics at the same level in mathematics and reading, correlation coefficients were obtained of which only a few were significantly different from zero.

The student variables to be studied in conjunction with this measure were those of intelligence and level of reading achievement. No general significant relationship was found in either the mathematics or reading curriculum between this rate measure and student intelligence. These results are consistent with the finding for the previous rate measure, number of days required to complete unit. The second variable, student level of reading achievement, also was found in general not to be associated with student average daily achievement in mathematics or reading and, again, this is in agreement with the findings of the previous measure of number of days to complete unit.

Relationship Between Two Measures

Of interest to this study is the relationship that exists
between the measures of student learning rate of number of
days to complete a unit, and average daily achievement. If
a relationship is found between the two measures of learning
rate, it would be anticipated that this would be a negative
one since the rate measure, number of days to complete a
unit, is the denominator for the rate measure of average
daily achievement.

The results indicate that when these two measures
are correlated, a moderately strong relationship exists in
units of mathematics while only a moderate relationship is
evident in units of reading.

Notes

1. John I. Goodlad and Robert H. Anderson. The Non-
    graded Elementary School. Revised. New York:
    Harcourt, 1963; p. 27-28.
2. Walter W. Cook and Theodore Clymer. "Acceleration
    and Retardation. " In Individualized Instruction, The
    Sixty-First Yearbook of the National Society for the
    Study of Education, Part 1 (edited by Nelson B.
    Henry).
3. Patrick Suppes. "Modern Learning Theory and the Ele-
    mentary School Curriculum. " American Educational
    Research Journal 1:79-94, 1964.
4. Robert Kalin. Development and Evaluation of a Pro-
    grammed Text in an Advanced Mathematical Topic for
    Intellectually Superior Fifth and Sixth Grade Pupils.
    Doctoral dissertation. Tallahassee: Florida State
    University, 1962.
5. Herbert Woodrow. "Interrelations of Measures of
    Learning. " The Journal of Psychology 10:47-73, 1940.
6. John B. Carroll. "A Model for Learning. " Teachers
    College Record 64:723-33, 1963.
7. R. L. Ebel. Measuring Educational Achievement.
    Englewood Cliffs, N. J.: Prentice Hall, 1963.
8. George L. Gropper and Gerald C. Kress. "Individualiz-
    ing Instruction Through Pacing Procedures. " AV
    Communication Review 13:166-68, 1965.
9. Herbert Woodrow. "Interrelations of Measures of
    Learning. " The Journal of Psychology 10:49-73, 1940.
10. Arthur Jensen. "Learning Ability in Retarded, Average,

and Gifted Children. "   In Educational Technology
(edited by John P. DeCecco).   New York: Holt,
Rinehart and Winston, 1964; p. 375.
11. Robert Glaser, James H. Reynolds, and Margaret C.
Fullick.   Programmed Instruction in the Intact Classroom.   Project 1343, Cooperative Research, US Office of Education, 1963.

EFFECT OF IPI PROGRAM IN ARITHMETIC ON PUPILS AT DIFFERENT ABILITY LEVELS

Donald Deep

In order to determine whether or not there were differences in the progress of different ability level students in
the Individually Prescribed Instruction (IPI) program the following data pertaining to the fourth, fifth, and sixth grades
at Oakleaf Elementary School were compiled:

(1)   Number of days each student attended school.
(2)   Number of skills a student completed or mastered in a school year.
(3)   Number of units a student mastered by doing assigned tasks in a school year.
(4)   Number of units mastered by a pretest score of
85 per cent or above in a school year.
(5)   Total units.

1.   Assumptions Underlying the IPI Program

The following assumptions were thought of to be guiding principles in an individualized instruction program:

(a)   One obvious way in which pupils differ is in the
amount of time and practice that it takes to master given instructional objectives.

(b)   One important aspect of providing for individual
differences is to arrange conditions so that each student can
work through the sequence of instructional units at his own
pace and with the amount of practice he needs.

(c) If a school has the proper types of study materials, elementary school pupils working in a tutorial environment which emphasizes self-learning can learn with a minimum amount of direct teacher instruction.

(d) In working through a sequence of instructional units, no pupil should be permitted to start work on a new unit until he has acquired a specified minimal degree of mastery of the material in the units identified as prerequisites to it.

(e) If pupils are to be permitted and encouraged to proceed at individual rates, it is important for both the individual pupil and for the teacher that the program provide for frequent evaluations of pupil progress which can provide a basis for the development of individual instructional prescriptions.

(f) Professional trained teachers are employing themselves most productively when they are performing such tasks as instructing individual pupils or small groups, diagnosing pupil needs, and planning instructional programs rather than carrying out such clerical duties as keeping records, scoring tests, etc. The efficiency and economy of a school program can be increased by employing clerical help to relieve teachers of non-teaching duties.

(g) Each pupil can assume more responsibility for planning and carrying out his own program of study than is permitted in most classrooms.

(h) Learning can be enhanced, both for the tutor and the one being tutored, if pupils are permitted to help one another in certain ways.

2. Materials in the IPI Program

The mathematics curriculum at Oakleaf was developed by first defining a sequence of behavioral objectives for grades K through six and beyond. The materials used were selected on the basis of these objectives. The statement of the behavioral objective indicated the desired change in the child whenever he successfully completed a designated skill.

Specific criteria were used as guidelines in writing the behavioral objectives. Lindvall[1] suggested three: the

objective should be stated in terms of the pupil; the objective
should be stated in terms of observable behavior; and the
statement of an objective should refer to the behavior or
process and to the specific content to which this is to be
applied.

After the objectives were identified, materials were
selected to teach each behavioral objective.   Much of the
material was self-study in nature, that is, materials that
a pupil could study or work by himself with a minimum of
teacher direction.   Worksheets from commercial arithmetic
books and other work pages were used to teach the behavioral objectives in mathematics.   Whenever a gap was located
in the material, Learning Research and Development Center
personnel or the teachers at Oakleaf would write work pages
designed to teach this skill.

Many gaps were also eliminated by specific teacher
instruction within the classroom.   Hundreds of commercial
pages were placed into specific units at varying levels of
difficulty with the intent of matching work pages with a behavioral objective.   There was no attempt to select any
series of textbooks, workbooks, or any related material that
would fit any particular curricular approach.

Although much of the material was self-study, that
did not mean that there was no teacher-pupil interaction.
Actually the total plan called for small group instruction,
large group instruction, and individual tutoring by the teacher.

3.   Procedures of the IPI Program

The amount of time spent on the (IPI) program involved about 45 minutes per day each for math and reading.
For the remainder of the school day the students engaged in
study under procedures followed in the other Baldwin-Whitehall schools.

An extensive placement testing program was initiated
during the first month of school.   This aided in placing a
student where his capabilities made it possible to successfully begin the program.   It was essential to the program
that the individual differences could be identified and a prescription of work could be developed for each child.   From
the placement tests, the youngster was assigned a unit-level

at which he began working.

The student was then given a diagnostic test (pretest) over the particular unit in order to determine the skills for which mastery or lack of mastery was indicated. If he fell below a particular score (85 per cent) on the pretest he had a prescription written for him which decided how much and what skills had to be assigned. These pages of materials had been previously identified as to the skill they were to teach and the behavioral objective they would accomplish. This might have been enough material for a day, several days, or a week depending upon the ability of the student and the difficulty of the work.

A student then began working on his prescribed materials, usually studying by himself. This type of individual study was done at a desk in a study area seating 80 or 90 pupils. These areas are designated as the intermediate or primary learning centers. The student went to carts in which the materials of the curriculum were placed and collected work pages to complete his own prescription. In attendance at the carts was a teacher aid whose purpose was to help the student find the materials quickly and keep a constant inventory of the supply of materials.

After filling his prescription the student returned to the instructional center and began working. In this center there were two or three teachers who provided instructional assistance and three or four clerks to distribute materials and grade papers. When a student completed an assigned task he took it to one of the clerks or teachers who scored it immediately and recorded it on the student's prescription progress sheet. If he mastered the material on the work page he moved on to the next task. If not, he checked or corrected his work or sought assistance from one of the teachers.

In this manner most pupils were able to proceed through their study materials with a minimum of help from the teacher. If a teacher found a pupil who needed more help than she could give him in this large group situation, the pupil was directed to a small room where another teacher gave him individual help or would involve him in small group instruction. The constant recording of the students' scores on material developed into a progress chart for the student and indicated whether or not the skill sheets were adequate. It also enabled a judgment to be made on what

the student should do next.   In the skill pages some were
identified as "curriculum embedded" check tests.   The student viewed these only as another work page but they helped
to determine more accurately whether or not he was ready
for a posttest or should continue doing more work.   After
carefully reviewing a student's work, a teacher decided when
it was time for the posttest in the unit in which the student
was working.

When the student had to take a posttest he went to
the test center, got his test, took it, and then returned it
to the test center.   There an aide scored the test and returned it along with all of the student's work in the unit.
With the help of this information the teacher decided the next
prescription.

Results of the Study

The data obtained in studying whether or not pupils
of different ability levels differ in their progress and achievement under IPI and how any such differences compare with
those found in schools using more conventional programs of
instruction are presented in seven sections:

(1)   Progress data for different ability students under
IPI.
(2)   Achievement in arithmetic computation by fourth-
grade students in the experimental and control groups.
(3)   Achievement in arithmetic problem solving by
fourth-grade students in the experimental and control groups.
(4)   Achievement in arithmetic computation by fifth-
grade students in the experimental and control groups.
(5)   Achievement in arithmetic problem solving by
fifth-grade students in the experimental and control groups.
(6)   Achievement in arithmetic computation by sixth-
grade students in the experimental and control groups.
(7)   Achievement in arithmetic problem solving by
sixth-grade students in the experimental and control groups.

Each of the sections, B through G, provides the variable means and the adjusted means for the high, average,
and low experimental and control groups.   Each section also
includes an analysis of covariance summary providing the
source of variation, sum of squares, degrees of freedom,
mean square, and the significance or non-significance of
F.

TABLE 1

Mean Values for Days Attendance, Number of Skills and Units Mastered, and Number of Units Pretested Out, for Fourth-, Fifth-, and Sixth-Grade Students Studying Under IPI

Ability Group	Days in Attendance	Skills Mastered	Units Mastered	Assessed Units Mastered	Total Units
4th High	181.06	25.33	9.00	6.22	15.22
Aver	179.85	20.60	6.90	6.30	13.20
Low	179.38	20.75	8.00	4.75	12.75
5th High	180.42	24.00	8.17	5.83	14.00
Aver	180.90	27.10	9.10	6.10	15.20
Low	180.40	19.60	7.40	5.40	12.80
6th High	181.71	33.43	9.14	7.14	16.29
Aver	184.1	36.8	9	7.4	16.40
Low	181.35	27.5	8.7	4.8	13.50

One basic type of data that is meaningful is examining the effectiveness of the IPI project with students of different ability levels is that pertaining to the amount of content covered during the school year (see section above).   Are there differences in the rate for mastering arithmetic skills of different ability level students?   Do brighter students master more units during a given year than average or slow students?   Do slow students show an appreciable amount of content mastered?   Data with respect to such questions are presented in Table 1, which gives mean scores for the fourth-, fifth-, and sixth-grade high, average, and low ability groups.

School was in session for 185 days in 1964-65.   The number of skills represents the number of behavioral objectives a student successfully mastered by working lesson pages which were designed to teach the skills.   The number of units involves those which a student successfully mastered by doing the necessary lesson pages for the many skills within a unit.   As was previously mentioned, a unit consisted of many skills.

In some cases a student did not have to work on any lesson pages in order to master a particular unit.   If the student received a high pretest score on a unit, the unit was considered mastered and the student was assigned work in another unit.   This prevented the overlap of teaching something previously learned by the student.   The last category in Table 1 ("Total Units") gives the sum of the units actually worked in and mastered and units which were mastered by a high pretest score.   In observing Table 1 it was expected that the high ability group in each grade would show mastery of a greater number of units than the average group and that the average group would have greater mastery than the low group.   Generally speaking, this tended to be true, but some inconsistency was revealed.

Findings

The study proposed to answer the following question: Is there a difference among pupils at different ability levels in their progress and achievement under IPI and how do any such differences compare to those found in schools using more conventional programs of instruction?

(1)   Progress Data for Pupils at Different Ability

Levels under IPI.

In general, the higher ability groups tended to do more work than the lower ability groups. Although there was some inconsistency within each grade, the numerical differences were not large enough to be of any practical significance. The most frequent inconsistency was that of the average group scoring better than the high group. However, the size of such differences was very small.

The progress data indicated that the higher ability students mastered more arithmetic skills and units than the lower ability students. It was hoped that the IPI program would not discriminate against pupils at any level of aptitude and that each of the ability groups would show definite progress in arithmetic. The data seemed to support this along with showing that the more able students did cover more material during the course of the school year.

(2) Achievement in Arithmetic Computation by Fourth Grade Students in the Experimental and Control Groups.

There was no significant difference among high, average, and low ability fourth-grade Oakleaf students in arithmetic computation scores whenever the pretest performance was taken into account. This is supportive evidence that the IPI program at Oakleaf Elementary School did not operate differentially for any ability group in the fourth grade.

A significant difference in the computation scores was reported in two of the control schools. The analysis of data for the third control school indicated no significant differences in the scores of its three ability groups.

(3) Achievement Scores in Arithmetic Problem Solving by Fourth-Grade Students in the Experimental and Control Groups.

There was no significant difference among high, average, and low ability fourth-grade Oakleaf students in the arithmetic problem solving scores whenever the pretest performance was taken into account. This is further supportive evidence that the IPI program did not differentially affect achievement of the three ability groups in the fourth grade.

The data for one of the control schools indicated a significant difference in the problem solving scores of the different ability fourth grade students, but the F values which were reported for the other schools indicated no significant difference.

(4) Achievement in Arithmetic Computation by Fifth-Grade Students in the Experimental and Control Groups.

There was no significant difference among high, average, and low ability fifth-grade Oakleaf students in the arithmetic computation scores whenever the pretest performance
was taken into account.   These findings correspond to those
found in the fourth-grade.

Once again the analysis of data for two of the control
schools did indicate significant differences in the computation
scores of their students under conventional instruction, and
the data for the third school indicated no significant difference in the scores for their fifth-grade students.

(5) Achievement in Arithmetic Problem Solving by
Fifth-Grade Students in the Experimental and Control Groups.

There was no significant difference among high, average, and low ability fifth-grade Oakleaf students in arithmetic problem solving scores whenever the pretest performance was taken into account.   This again corresponds to the
findings reported for the fourth-grade students at Oakleaf.
This evidence suggests that when achievement is measured
by a standardized test the IPI program has a similar impact
on the three ability levels: high, average, and low.

The data for each of the three control schools also
indicated no significant difference in the problem solving
scores of their ability groups.

(6) Achievement in Arithmetic Computation by Sixth-
Grade Students in the Experimental and Control Groups.

There was no significant difference among high, average, and low ability sixth-grade Oakleaf students in arithmetic computation scores whenever the pretest performance
was taken into account.   This result is comparable to that
found for the fourth and fifth grades at Oakleaf.

The data for each of the three control schools indicated no significant differences.   All schools, experimental
and control, indicated no significant differences among the
fifth-grade ability groups using problem solving scores and
among the sixth-grade ability groups using computation
scores.

(7) Achievement in Arithmetic Problem Solving by
Sixth-Grade Students in the Experimental and Control Groups.

There was no significant difference among high, average, and low ability sixth-grade Oakleaf students in the
arithmetic problem solving scores whenever the pretest performance was taken into account.   All three grades under
IPI thus reported no significant difference among their ability
groups whenever computation or problem solving scores were

used in the analysis.

The data for one school indicated a significant difference in the problem solving scores for its sixth-grade students but the data for the other two schools reported $\underline{F}$ values which were not significant.

Conclusions

The results of progress data for different ability students under IPI indicated, in general, that higher ability
students mastered more skills and units than did the lower
ability students.   However, despite the suggestion that the
bright students progress faster and master more material
than the slower students, the results of the standardized
tests used in the study seem to raise a contradiction. These
results indicated no significant difference among high, average, and low ability students in arithmetic computation or
problem solving scores whenever the pretest performance
was taken into account.

A possible answer to the contradiction may be that
standardized tests measure content which is appropriate for
a particular grade level but inappropriate for measuring
achievement in the IPI program.   For example, the elementary battery which is given to third- and fourth-grade students measures content material suitable for the third and
fourth grades.   In the IPI program, however, a fourth-grade
student is not limited to performing arithmetic tasks only on
the fourth-grade level.   A new student entering the IPI program is given a series of tests in order to determine his
starting place in the math continuum.   If it is thought best
for the individual, a fourth-grade student may be allowed to
work in second-, third-, fourth-, fifth-, and even sixth-
grade arithmetic skills.   Also a sixth-grade student may be
working with fourth-grade skills in fractions during one week
and fifth-grade skills in addition the following week.   Once
mastery is indicated, the student is then permitted to do
work at a higher level.   The only limitation placed upon the
student in how fast he can proceed within the IPI program
is his ability to learn.   It seems very possible that under
IPI the students are gaining mastery of skills that fall outside the range of those covered by the given standardized
test.   That is, some may be mastering abilities that could
be assessed only by giving a higher level test while some
are mastering content of a lower grade level that could only
be evaluated by a lower level test.

In the comparison of the experimental and control schools, it did appear that the instruction in two of the control schools may have operated differentially in some instances on the different ability groups.   That is, in these schools the standardized test results did show that there was a difference in the gains shown by students at different ability levels.   This might well be anticipated in situations where all pupils spend the school year in a concentrated study of the content of one given grade level.

The third control school had results similar to those of Oakleaf: in either the fourth, fifth, or sixth grade and with computation or problem solving scores, there was no significant difference reported among high, average, and low ability students.

Note

[1]C. M. Lindvall.   Testing and Evaluation: An Introduction. New York: Harcourt, Brace and World, 1961; p. 23-25.

VIII.   RESEARCH: 2.   Self-Selection

I am certain that I have not concealed my conviction that the principles of self-selection are the keystone of any sound, viable plan of individualized instruction.   The two parts of the current chapter are excerpted from doctoral dissertations in which the viability of self-selection was the principal subject of the reported investigations.

William T. Ebeid

The first of these theses was completed at the University of Michigan in 1964.   Its author, Dr. Ebeid, is presently a professor at Assuit Teachers College in the United Arab Republic.   I consider this one of the half dozen or so best doctoral theses that I have read.   One can find in it ample evidence of persistence and a genuine scholarly approach to the solution of the problem chosen.

It is obvious that individualized instruction is viable only if there are major individual differences among learners. Dr. Ebeid therefore very logically began with a step-by-step investigation of individual differences.   His report of that study, which is extraordinarily well documented, is well-done and will repay the reader's careful examination.   I have included a substantial portion of this part of his thesis, which seems to me to present an unanswerable argument for the necessity of recognizing and coping with individual differences in the educational process.

Following this portion of his thesis, Ebeid turns to an equally careful analysis of the innumerable ways in which attempts have been and are being made to accommodate to individual differences in educational processes.   His exposition of the material gathered by him is in my opinion highly informative and undoubtedly is one of the best compilations of the methods of coping with individual differences in

education.   It is carefully documented as well.   To my re-
gret, space simply is not available to present any substantial
excerpt from this part of his material.

An excerpt of the actual study performed by Dr.
Ebeid is also included.   The excerpt is much too short to
do the original full justice, but I believe its quality speaks
for itself.   Because of the brevity of the part of the excerpt
included in this chapter, I am quoting here from the dissertation written by Professor K. Allen Neufeld, which contains
a concise description of the Ebeid experiment.   While this
may possibly be somewhat redundant, I think it important
to assure that Ebeid's principal study will be seen in proper
perspective.   Excerpts from Professor Neufeld's thesis were
included in a previous chapter.   The summary reads as follows:

Ebeid conducted a study in arithmetic with 47 seventh-
graders and 47 eighth-graders as his experimental
group.   The control group consisted of 75 seventh-
graders and 88 eighth-graders.   The pupils represented a wide range of intelligence levels.   The con-
trols used SMSG text materials and conventional instructional procedures were utilized.   The experimental
group were subjected to individualized instruction for
20 per cent of the time during the first semester and
40 per cent during the second semester.   A variety
of mathematical materials were used.   During the remainder of the time, the instructional method paral-
leled that of the control group.   The individualized
learning activities were analyzed in five categories.
Cultural activities consisted of puzzles, games, history, and function.   Manual activities included the use
of adding machines, the abacus, and devices such as
the Towers of Hanoi.   Textual activities were basically
units of content from the regular SMSG materials
which a student might self-select and work on at a
self-chosen pace.   Conventional activities were classified as units of content selected from textbooks other
than the regular class text (SMSG).   Explorational
activities involved units of work from advanced textbooks.   Thus the individualized instruction provided
both enrichment and acceleration.
Using analysis of covariance with posttest achievement as the independent variable and pretest achievement and/or intelligence as covariates, the following
findings were reported.   There were no significant

differences between the experimental and control students in their mathematical achievement as measured
by both STEP and SMSG tests. Although the differences were not statistically significant, the experimental students averaged higher than the control on the
STEP achievement tests. Ebeid further reported that
'Although the students recognized that this self-selection approach increased their knowledge in and about
mathematics, yet they stressed more its psychological
contributions. They frequently expressed an appreciation of the opportunity to work independently, each
at his own level of interest, rate, and ability. '

Henry D. Snyder, Jr.

     The second part of this chapter is excerpted from
another University of Michigan doctoral thesis. It is, incidentally, not by chance that so many University of Michigan theses are concerned with one form of individualization
or another as that institution has long been in the forefront
in the development and dissemination of ideas about individualization of instruction.

     Dr. Snyder, who in his thesis describes a one-year
experience with a self-selective mathematics program at the
junior high school level, is a member of the Education Department of the University of Wisconsin in Milwaukee. During the time that he has been there he has been in charge
of the kindergarten through 12th-grade mathematics teacher
education program. After five years of this kind of activity,
Professor Snyder felt that his "awareness of real school situations [was] becoming cloudy. " This is a dilemma which I
think occurs in the life of everyone who finds himself involved in teacher education in the rapidly changing situations
that are typical of the times in which we live. Most of us,
especially including myself, notice this dilemma, deplore it,
feel that something should be done about it, proceed to follow our usually busy programs and routines, and eventually
adjust ourselves to ignoring the dilemma insofar as it applies to us. Not so Dr. Snyder! He made arrangements
with the Milwaukee Public Schools and with the university
to allow him to use half his university program to spend
full days and teach three classes at Lincoln High School in
Milwaukee. My hat is off to someone who is not only a
recognizer of dilemmas, as I am, but who is also one who
makes an action-oriented decision about his predicament. It

is hard to disagree with his statement that "my university classes have got to be more realistic after this. "

I will not undertake to do more than let Dr. Snyder's selection in the second part of this chapter speak for itself.

SELF-SELECTION AND THE TEACHING OF MATHEMATICS

William T. Ebeid

Objectives of the Study

(1)   To determine the relative effectiveness of the experimental approach as compared to the conventional approach to mathematics instruction, namely, to the approach which devotes all its time to group instruction and the use of a single textbook.   This comparison will be in terms of the students' mathematical achievement; and the students' attitudes toward mathematics.

(2)   To study the dynamics of the learning situation in which the experimental students used the self-selection principle for their mathematical learning.   This includes:

(a)   A presentation of the procedures of the experiment;

(b)   A study of the learning activities in which the students were involved in the self-selection periods;

(c)   A study of the students' self-selection ability, as will be defined later, and its relationship to a number of the students' other measurable attributes;

(d)   A study of the reactions of both the students and the teachers involved to the different aspects of the experiment.

Methods of Research

A control group was used, the subjects which were four seventh-grade classes and four eighth-grade classes. The mathematics program for the control group was full-time group instruction within a framework of most conventional classes.   In other words, the same textbook, sequence and

group instruction were used for the weekly five periods which
constituted all the time allotted to mathematics instruction.
The two groups, experimental and control, were compared
in terms of mathematical achievement and mathematical attitudes, using two instruments.

The California Short-Form Test of Mental Maturity
and the Mathematics Sequential Tests of Educational Progress (STEP). This study intends to test the following hypotheses:

H1.   There are no significant differences between the exper-
       imental groups and the control groups in their general
       mathematics achievement as measured by the STEP
       tests.
H2.   There are no significant differences between the exper-
       imental groups and the control groups in their textual
       mathematics achievement as measured by the SMSG
       Mathematics Inventory.
H3.   There are no significant differences between the exper-
       imental groups and the control groups in their attitudes
       towards mathematics as measured by the Mathematics
       Attitude Scale.

1.   Individual Differences in Human Thought

"We broach here a new subject, difficult and as yet
very meagerly explored."[1] With these words, Binet and
Henri started their article of 1895 about individual differences. At that time the study of individual differences was
becoming a new science. Through the history of human
thought, these differences and their possible significance
have been investigated through speculation, introspection, observation, and laboratory techniques. Differences have been
shown to exist in physical size and shape, physiological
functions, motor capacities, sensory and perceptual sensitivity, intelligence, interest, attitudes and personality traits.

Biologists present what they consider clear-cut evidence for human uniqueness. Medawar[2] in his book, The
Uniqueness of the Individual, cites the example of skin graft-
ing: donated skin is rejected unless an identical twin sibling
is the donor. Garret Harden[3] presents 14 biological characteristics which illustrate a wide range of variation.

The psychologist Bessel, who discovered that there

was considerable variation among individuals in the speed with which they reacted to a visual stimulus, demonstrated the possibility of numerical description of the way an individual's nervous system functions.   Quetelet, the Belgian mathematician, applied the mathematical theory of probability to human measurements.   Francis Galton developed ingenious methods for determining the degree of sensitivity and strength of an individual's imagery.   W. Wundt shifted from being interested in discovering general laws of human nature to being interested in the difference between his subjects, and the possible significance of these differences.   McKeen Cattel initiated the mental test movement which has become increasingly important from 1890 to the present time.   A whole team of psychologists measured all aspects of sensation, perception, attention, discrimination, memory, association, and speed of reaction of the individual.   Such factors were measured to evaluate the total efficiency of the individual, assumed to be an index of the individual's general intelligence.   This assumption was based on the theory of the time which considered that all of mental life was built up of units of sensory experience, just as all the physical world is made up of atoms.

Progress in the mental test movement came with the work of Alfred Binet and Simons in 1905. [4]   Binet held that the complex mental abilities we classify under the term intelligence are not made up of single abilities.   He introduced the concept of mental age which has been supplemented by Stern and Terman's concept of Intelligence Quotient (IQ = mental age divided by chronological age) to get an index of the rate at which mental growth occurs in a given individual.

In general, we can summarize the testing movement in the following phases of development:

(1)   The search for group tests.   This became important during World War I with Otis' work. [5]
(2)   The development of non-verbal tests of intelligence.   This was to find tests independent of school knowledge.   There were both group and individual tests.   They are still used as supplements and for clinical studies.
(3)   The concern about the different aspects of intelligence such as spatial judgment, numerical ability, verbal ability, etc.   Along with that phase came the development of special tests for different levels rather than the search for universal intelligence tests suitable for all human beings under all circumstances.

(4)   The development of tests for specific talents and
vocational aptitudes.   This helped to produce tools useful in
selection, placement, and individual guidance.
(5)   The search for precise evaluation of the non-in-
tellectual traits of the individual, such as interests, adjustment patterns and personality traits.   E. Strong's efforts
and contribution in this work are still dominating. [6]   The
projective technique made its appearance since the 1930's as
a new approach to appraise such traits as attitudes and
drives.   Projective techniques are not quantitative, but they
are very useful in clinical studies.
(6)   The use of objective laboratory techniques to investigate differences in personality and temperament.   In
this line of development emphasis is on accurately measurable perceptual or motor responses rather than on such self-
report techniques as personality inventories.

2.   Individual Differences and Child Development

The relation between educational practices and child
psychology is deeply rooted in the history of education.   The
Egyptian scribe used to treat his children as young adults,
believing that the child began as a minute man, the homun-
culus, who grew larger during gestation and then was born
when he reached a certain size.   Even the way we look at
the child in school has considerably changed along with change
in our knowledge of child development, yet our educational
practices are still "fixed" in establishing age norms for
specific aspects of children's development.   We used to find
in the usual textbook a chapter on norms of motor development, sensory development, mental development, etc. , which
reports the findings on specific samples of children who have
been measured for one variable that is then correlated with
chronological age.   Consequently, educational practices had,
and still have, their programs of courses distributed in
terms of grades, which are solely a function of chronological
age.

Curriculum organization, then, has ultimately been
based on the norms and similarities that exist among children of the same age.   Recent knowledge of child development has moved beyond the "norms" phase of study, taking
into consideration growth phenomena as a whole by trying to
fit together information on the many facets of a child's life.
The focus has become the individual and the preferred practice has become to use the entire functioning child as a

laboratory subject.   However, we are at a stage where educational practices have not yet caught up with our knowledge
of the ways children grow.

The study of child development is concerned with the
nature and regulation of the changes in structure, function,
and behavior which occur in children as they progress from
fetal infancy through adulthood.   Such studies have revealed
and identified orderly stages of growth, [7] dividing it into
arbitrary periods according to chronological age and pointing
out characteristics common to each group in such periods.
Deep analysis, however, has revealed overlapping of these
age periods, pointing to the existence of individual differences in development rates.

In a review of the characteristics of four-year-olds,
Minnie P. Berson states, ''There is probably no single child
who evidences all the typical characteristics of the four-
year-old as described by averages derived from studies of
populations of that age. ''[8]   Berson points out an important
fact in child growth in the statement ''Composite charts show
similarities among children.   But they also show ranges and
variability in growth and maturation. ''   Harold Jones and his
associate presented individual profiles of 14-year-olds
(eighth graders) showing their striking differences in size,
personal and interpersonal security, motivations, defenses
and attitudes. [9]   Lawrence Frank expresses the uniqueness
of children in these words:

> Studies of newborns, infants, toddlers and preschool chil-
> dren suggest that there is an orderly sequence of growth
> and development from conception on, like a broad highway
> along which each child with his unique heredity, his indi-
> vidual nutrition and nurture and life experiences will
> move .. at his or her own rate of progress, attaining
> the size, shape, functional capacities, and abilities that
> are uniquely his or her own.   This applies also to adoles-
> cent growth and development. [10]

When growth in its connotation of increase in size has been
studied, changes in some aspects appeared to occur with a
regularity which might give prediction some accuracy.   This
is exemplified by change in height and weight as related to
age and in accordance with body structure.   But, even in
physical measurements, exact predictions for the future of
an individual are made cautiously and with a realization that
unknown variables may affect status and progress. [11]

Similar studies are found in Gesell's work, which shows a relationship between change in motor development and age.[12] Stages through which many children proceed have been identified in sensory development (Zubeck and Sobbery 1954), language (McCarthy 1954), cognition (Gshelder and Matalon 1960), creativity (Alschuler and Hattwide 1947), emotional development (Jersild 1954) and social development (Bridge 1931).

Courtis distinguished four general cycles of growth: prenatal, infancy, childhood and adolescence, claiming that it is possible to describe all growth that occurs in a particular cycle by a universal growth curve.[13] However, Olson reported that the application of such a master curve is complicated in practice, because it is found that individuals in a group are at different stages in a cycle because of differential growth rates and that each is developing to a different final status.[14] Arnold Gesell, after clinical study of thousands of behavior patterns and pattern phases at 34 age levels from the period of fetal infancy through the first ten years of life, concluded that no two individuals are alike even though they follow a general ground plan of growth. He considered such a general ground plan of growth as a characteristic of a cultural group within the species--the same reasoning which Cattel followed when he attributed many of our similarities to what he called "environmental mold unities."[15]

Gesell emphasized that "It can be said with scientific assurance that when the infant enters the world, he is already an individual with growth potentialities which are distinctly his own." Beginning with infancy and with the help of cinema, Gesell identified numerous similarities and differences which demonstrate the basic role of maturation in shaping the individuality of physical and psychical behavior even in extremely similar twins.

Dorothy Eichorn, in a review concerning the biological correlates of behavior, reports that "much of human behavior is a function, not of structure per se, but of the environmental supports, outlets, expectations, and reactions relative to that structure."[16] She attributes differences in interests and behavior also to varying thresholds for some classes of stimuli.

While individual differences are customarily expressed in perceptual, emotional, cognitive, social, and motor behavior, we find some reports like that of Kagan, Moss and

Sigel which assert a tremendous distinction among subjects which cuts across many traditional categories. [17] This made White think of what he called a new dimension in individual differences: at the root, they--categorized differences--appear to be variations in impulsiveness, attentiveness, and the ability to focus closely on the environment. [18]

   Interest in putting the child together and studying the interactions of the many facets of his life began early in the child development movement. Each child is observed either for part or the whole of his developmental period, as exemplified by the longitudinal studies of the University of Michigan, the Brush Foundation, the Fels Research Institute, the Child Research Council, the Harvard School of Public Health, and the Yale Clinic of Child Development. Gesell's studies in 1948 emphasized that "The unity of the organism is not conceived in mystical terms, but is identified with the functional and structural unity of the total reaction pattern. "

   Stress upon the total growth of the child and the interrelatedness among the many variables of his growth tend to be the fashion in the recent literature to the extent that many "tautological" concepts appear frequently. References to the "constitution" of the child, the "system" of the child, the "genotype" of each individual, growth age, educational age, organismic age, etc. , are familiar to students of child development. The "constitution" of the child, to Witmer and Kotinsky is "the sum total of the structural, functional and psychological characters of the organism. "[19] Sigel thinks of the "whole child" as an organization of a number of systems, which function at varying degrees of autonomy and interrelatedness. [20]

   In the context of the "organismic" conception, Olson asserts that changes of various structures and functions of the growing child are interrelated in some type of dynamic balance. In computing the "organismic age, " he averages the child's mental age, reading age, dental age, height age, weight age, grip age, and metacarpal age. The organismic age appears to be heavily weighted in the direction of physical development.

   The uniqueness of the child and the orderliness of his growth do not negate differences within aspects of growth of the same child. The intercorrelations among these variables are flexible enough to permit differentiation and individualization of some specific aspects from the total pattern. In

other words, "although various abilities overlap, that is, intercorrelate, they are sufficiently independent to merit separate measurement."[21] One recalls here Olson's statement
that "the child has many ages."

The study of child development gives us new dimensions in our knownledge about motives and cognitive development in children. White in his review of learning asserts
that "Recently psychology has paid new attention to motives
associated with curiosity, stimulation-seeking, and exploration." Children get satisfaction from manipulations and
mastery of their environment. Richness of the environment,
novelty and challenge constitute intrinsic motivation to children. As they grow, their acceptance of the challenge of
a task grows and objective self-evaluation dilutes the value
of praise.[22]

Hunt and Fowler urge an early realization of potential
abilities. Hunt has summarized biological and psychological
evidence which leads him to question the assumption that intelligence is fixed and its development predetermined by
genes. This emphasizes the value of an enriched environment for intellectual development. On the basis of evidence
from animal learning, neuropsychology, computer programming, and Piaget's studies of intelligence in the child, Hunt
recommends that "the early environment of children should
be more intellectually stimulating," resulting, he believes,
in a higher level of intellectual capacity. Fowler, like Hunt,
urges that children in the early years be offered more cognitive stimulation.[23]

We can summarize the following theses to furnish a
psychological background for the educational project of this
study:

(1) Individual differences exist in all aspects of human functions and structures.
(2) Individual differences exist at any chronological
age and at any of the recognized growth periods.
(3) Growth is continuous in all traits and the approach
to maturity is gradual.
(4) There is an orderly sequence of growth and development from conception on. But equally significant is
that each child, with his unique heredity, his individual maturation and nurture, and life experience, will grow at his
own rate of progress, attaining the size, shape, functional
capacities, and abilities that are uniquely his own.

(5)   Each child has a unique pattern of growth but that pattern is a variant of a basic ground plan.

(6)   There are differences in the traits within the same individual.   Hence a child has many ages at a certain time.

(7)   The various special abilities correlate differentially with each other and with any measure of general ability that is presently used.

(8)   Differences are not static.   Children vary when they enter school and they develop at different rates.   Consequently diagnosis must be a continuous process.

(9)   Differences not accounted for by mental age measures are as important in learning as intellectual ones.

(10)   There is a possibility of optimum development for all within the frame of growth.

(11)   All persons are able to make gains in the direction of desirable personal and social goals, even though all may not attain the same eventual level.

(12)   The idea of standards applicable to populations in general has given way to evidence that the individual child is his own standard.

(13)   The uniqueness of the individual and the existence of individual differences do not deny that there are central tendencies which should serve but as a general guide and as an approximate indication in approaching problems of certain age.   Meanwhile, deviation from norms is "normal."

(14)   It is difficult to change growth patterns in a substantial fashion by conscious environmental manipulations.

(15)   Much of human behavior is a function, not of his own physical-mental structure per se, but of the environmental supports, outlets, expectations, and reactions relative to that structure.

(16)   The early environment of children can be enriched by more intellectual stimulation.

(17)   The child can be intrinsically motivated through curiosity, stimulation-seeking, exploration, challenge, self-evaluation and feedback aspects of rewards.

(18)   The healthy child seeks from his environment those experiences that are consistent with his maturity and his needs.   We can trust this seeking behavior to tell us much about the readiness of a child for an experience.

Educational Implications

Such knowledge of individuality has presented a dilemma to the society as a whole, because of its challenge to the

profession of education.   Studies of learning took a clinical
turn as psychologists sought to determine what makes the
learner react as he does, and studies of learning curves de-
clined in popularity as experts found that these curves dealt
with averages instead of individuals.   Current literature in
learning contains more references to case studies of individuals and to analyses of factors influencing the course of
learning than to the refinement of the curves of learning.

On the teaching scene we find a gradual shift in attitudes towards individual differences from suppression to
tolerance to cultivation.   Without loss of generality, we can
say that most of the services offered to the idea of individual difference have been lip service.   Practices in the classroom, in most subjects like mathematics, are still the same
in terms of the same content for the whole group.   The idea
that concepts must be introduced and mastered en masse
persists.

There have been many attempts to improve curriculum
organization and presentation, to provide for individual differences, but these attempts have resulted only in minor adjustments.   Conflict results have also been reported in many instances, even with these minor adjustments.   In Wingo's
words, "The literature indicates the willingness of teachers
and administrators to accept the fact of individual differences
but uncertainty as to how the problem can be handled most
successfully."[24]

We can say that in general, and at all levels, education seems to be a conservative profession; we do not expect
sudden or drastic changes in its organization and procedures.
A schism between what educators know and what they do will
remains.   The state of uncertainity of how to meet individual
differences in the classroom comes basically from the absence of a distinct theory in education which can substitute
for the predominant conservative one.   The need for what
we can call a differential theory in education seems a priori
in such a time as this: educational theory must be in harmony with the ways children grow.

A Differential Theory

From the tremendous amount of materials written
about individual differences and from the literature related
to the use of self-selection in reading, it is possible to

formulate the following scheme for a differential theory for education:

A. Scientific Foundations   Knowledge of individual differences in general and the way children grow, as mentioned and summarized in the last section, acts as the scientific foundation for this educational theory.

B. Assumptions   (1) Education is a process by which children are assisted in their total growth; adequacy of its administration, physical environment, curriculum experiences, and methods of teaching should be appraised in terms of the extent to which this function is realized.   (2) Education must be geared to the needs of each child as a whole; this means that not only intellectual but physical, emotional, and social development as well must be considered.   (3) Studies do not yield unique chronological ages when all children can or should attain a certain level of growth; this is also true for factors of personality, socialization, physique or general intellect.   (4) Educational growth is attained optimistically in the frame of taking each child as he is and helping him to grow, step by step, in a direction and a process which is socially desirable and personally satisfying.   (5) Emphasis must be placed upon pacing the child instead of forcing him.   (6) Allowing a child to work at his own pace in what he sees as being of interest to himself helps him to develop his abilities on a realistic basis in terms of his potentialities and his level of aspiration; self realization and autonomous learning help the learner to tolerate the frustration that results from the search to make progress when encountered by difficulties. (7) There is a possibility of optimum development for every student; all persons are able to make gains in the direction of desirable personal and social goals, even though all may not attain the same eventual level.   (8) There is no essential conflict between wise nurture of growth and the other purposes of education such as full living in the present, preparation for the future, and community improvement.

A Classroom Model for a Differential Theory:
Self-Selection

The closest model for such a theory is based on the self-selection principle, in essence minimizing the need for systematic instruction, and allowing the child to select his learning materials according to his own interest and to

proceed according to his own rate.[25]   There are four major
dimensions to this model.

(1)   The Child.   The child has a natural tendency to
seek from his environment those experiences that are consistent with his maturity and needs.   Seeking behavior tells
much about the readiness of a child for an experience.   Pe-
diatric, psychological, and psychiatric considerations have
led to respect for self-regulation in bodily functions and to
a growth philosophy in matters of discipline and training.
It is known that throughout nature there is a tendency for
plant or animal life to be sustained by the self-selection of
an environment appropriate to its needs.   If such an environment does not exist or is inadequate, the human being works
creatively for the conditions that advance his well being.
Behind the tendency of a human being to self-select, there
exists a psychology of motivation which recognizes curiosity,
challenge, goal attainment, success, control of environment,
level of aspirations, self-evaluations and self-image as motives.   In the classroom, self-selection may be used as a
means of bringing together the nurturing qualities of learning
materials with the seeking tendencies of children.

(2)   The Environment: A Growth Medium.   Self-selection in the classroom implies the existence of a rich and
stimulating educational environment.   There should be many
alternatives available within different levels and different
aspects of a given subject, as well as different subjects.
The development of such a diversified environment can be a
cooperative enterprise among students, teacher, and librari-
an.

(3)   The Teacher.   The teacher's role in this situation is to help pace the child.   He has to ensure that each
child is provided with the materials upon which he can
thrive.   He also has to have a good assessment of each individual using the available anecdotal and quantitative records.   He must maintain an atmosphere of learning.   Feeling that success is within the grasp of the child is an example of the intrinsic motivations on which the teacher depends
in this situation.

(4)   Classroom Practices.   After preparing the classroom with such an environment, children, under the teacher's guidance, will be allowed to browse among and select
appropriate learning experiences.   The principle of self-
selection has been applied extensively in teaching reading in

the early grades and the results show satisfaction with this
principle in action.   The same principle has been applied by
pediatricians who allowed the child to select his own diet.
Our concern now is to find if it is possible to apply this
model in teaching mathematics as it is used in teaching reading.

With knowledge, we can bring both child-need and
subject-need together in a modified model of self-selection.
This is suggested to be through a skewed distribution of the
time allotted to the instruction of mathematics at any grade
between systematic teaching and self-selection learning.
This synthesis can be summarized in the following:

(1)   There is a need for a minimum structured core
of a mathematical knowledge and understanding.   This is going to occupy only a part of the mathematics program for
any given student at any given grade.   This core program
must be well-articulated, taking advantage of whatever light
modern psychology can throw on intellectual development and
the learning process.
(2)   There is a need for a wise use of the self-selection principle in mathematical learning.   By giving the student a structured contact with mathematics in the first part,
we can trust him to tell us about his tendencies, abilities,
and interests.   Therefore, the other part of the mathematics
program must absolutely be determined by the individual student.
(3)   The two parts of the program must be on a continuum in terms of each individual.   In the self-selection
part, the student can build on concepts he had in the structured part, make up for his weaknesses, accelerate his
learning, or even choose a totally different line from the
diversity of prepared mathematical materials.
(4)   The percentage of time devoted to each part of
this model must be flexible and in harmony with each situation.   And (5) this synthetical approach is assumed to fit
each student.   Thus its application must be for heterogeneous
groups.

Individual Differences in High School Mathematics

Most of the provisions to meet individual differences
problems in high school mathematics are concentrated around
three practices: ability grouping, acceleration, and enrichment, all of which are characterized by at least these

generalizations:

    (1)   There is no agreement on their effectiveness.

    (2) They are mostly oriented towards limited goals--
the search for raising the achievement of certain groups of
students--and almost no attempts have been made to investigate the educational process itself.

    (3)   Under most of the provisions, instruction is pre-
dominantly of a group all of whom have the same textbook
and the same sequence all the time.

    (4)   Other devices which appear more individualized,
such as enrichment, remedial work and acceleration, are
privileges given only to exceptional students, which deprives
the so-called average pupils.

    (5)   There is an indication that enrichment materials
can serve as motivating factors and that enrichment as a
device can be applied to students of varying abilities.

    Although in the field of mathematics there have been
few reports about this principle, in 1945 a minor but pioneer
self-selection study on the teaching of third- and fourth-
grade arithmetic was reported by Bernice Ault. [26] Without
the definition of a certain content or sequence of topics by
the teacher, the children gained more in arithmetic achievement.   On a more limited basis, Crosley and Fremont[27]
adopted the same technique in high school algebra where "no
particular sequence was adhered to.   However, students have
to select topics from the New York State Syllabus. " It was
also possible for related topics to be selected outside the
required syllabus with the approval of the teacher.   Brannon[28] reported about a course named "advanced mathematics" given in Urbana High School which utilized some self-
selection techniques.   Unfortunately, the course is open only
to students of high ability who could have completed three
semesters work in one year.   Every student of this class
works at his own level and chooses his own work in whatever sequence he likes but within the limits of the textbook
approved for the regular classes in the school.   Favorable,
but mostly informal, results are reported in such studies.

Results

    Before the experiment started the participating teach-
ers and the writer agreed:

    (1)   To use the same materials for the self-selection

periods for all the experimental classes.

(2)   To allocate to self-selection activities 20 per
cent of the total time allotted to the mathematics instruction
for the first semester.   This was defined operationally to
be two consecutive days each two weeks for each of the four
classes.

(3)   To use a calendar schedule for the self-selection
periods.   This was to secure better coordination in the use
of the materials and to permit the writer to observe, test,
and collect the appropriate data from all the self-selection
periods for all the involved classes.

(4)   To continue these meetings occasionally to discuss the proceedings of the experiment and its development.

In the first self-selection period the teacher of each
class explained to his students that each is allowed to select
from those provided any mathematical material he wants according to the level of his own interest and ability.   The
students were also allowed to select their textbook if they
wanted to do so.   "Laissez faire" appeared to be dominant.
The myriad differences among the students was reflected in
the many ways they coped with this new "freedom of study,"
as some of the students called it.   Mathematical games attracted some students, mere readings were not infrequent,
the textbook was the choice of at least one student in each
of the two seventh-grade classes, but none of the eighth
graders chose the textbook on his first self-selection day.
There were many students in each class who looked serious
and enthusiastic, others who tried to "goof around."   Others
appeared uncertain about what to do and some of them asked
the teacher to find something for them to do.   There were
tendencies for small "clans" to form, particularly in the
eighth grade.   Still others appeared to do more socializing
than learning.   It was easy to distinguish between introverts
and extroverts, dependent and independent learners, serious
and clowning students.   The first self-selection day in each
group was on the whole a clear display of the spectrum of
individual differences latent in the group.

As the experiment proceeded the students showed
more responsibility.   Even the "clans" took the shape of
small-group projects such as the eighth-grade one that
studied navigation with one of the student teachers, and the
two boys who spent a large percentage of their self-selection
time working together on stocks and the stock market.   At
the beginning of each period, each student picked by himself
some material to work on.   There were differences in the

kind of materials selected as well as obvious variances in
the depth, continuity and quality of the work undertaken.
There were times at which some students showed symptoms
of frustration, boredom, or inability to cope.   In such
cases, the teacher offered help.   (Student teachers and the
writer offered similar help.)   At the end of each period,
students were asked to report their activities and their comments on a free response basis and occasionally were asked
to write their reactions to the self-selection method in general.

In one sense, the first semester was taken as a pilot
study in which appropriate instruments were developed for
studying self-selection.   From the reactions of the students
and teachers, and with the consent and advice of the doctoral
committee, the percentage of time for self-selection learning
was increased to 40 per cent of the total time of the mathematics program of the four classes in the second semester,
as two consecutive periods each week.   Hence, data reported in this article are concerned with the second semester self-selection activities.

Q1.   How was the students' time in self-selection periods
      distributed among the different learning activities avail-
      able to them in these periods?
        The instruments used to answer this question were
the students' diaries, from which the reported mathematical
activities were surveyed and classified into cultural, manual,
textual, conventional and explorational activities, defined as:
CULTURAL: any activity which involved recreational or
reading materials as its major aspect--this included puzzles,
games, riddles, history of mathematics, biographies of
mathematicians, mathematical fiction, etc.
MANUAL: any activity which did not need much intellectual
sophistication in its performance--this included mere manipulation of computing devices such as adding machines, slide
rules, abacus, geometrical constructions, curve stitching,
surveying, etc.
TEXTUAL: any work related to the regular SMSG textbook
or to the systematic instruction part of their program such
as homework or classwork.
CONVENTIONAL: this included any work related to the conventional mathematics taught in regular classes such as
units on arithmetic, algebra, geometry, and general mathematics.   Remedial and accelerated work were listed under
this category.
EXPLORATIONAL: this included any unusual topic considered

more advanced than the above category--topology, pi, cal-
culus, etc.

These activities were tallied for each experimental
group.   The frequency of each activity was taken as an index of the total time spent in that activity.

Table 1 shows the time distribution for all the experimental groups in terms of percentage.   F and M stand
for females and males in the table.   Table 1 also shows
that the self-selection time was spent in the following decreasing order among the different activities: seventh
graders--cultural, conventional, textual, manual, and then
explorational; eighth graders--conventional, cultural, manual,
explorational, and then textual.

We also notice that in seventh grade, the time spent
on cultural activities is very close to the time spent on
conventional ones.   The same phenomenon appears with their
manual and explorational activities.   This does not appear
in the eighth-grade time distribution.   On the other hand,
when groups are considered in terms of females and males,
the table shows the following distribution--in a decreasing
order.

Seventh females: Cultural, textual, conventional, explora-
                   tional, manual
Seventh males: Conventional, cultural, textual, manual,
                   then explorational
Eighth females: Conventional, cultural, manual, textual,
                   then explorational
Eighth males: Cultural, conventional, manual, explora-
                   tional, textual

The table and the information taken from it reflect
the following trends:

a)   Both cultural and conventional activities came
ahead of others in the students' selections.   However, the
seventh graders spent more time than the eighth graders
did on cultural materials.   On the other hand, the eighth
graders spent more time on conventional activities.   This
might be related to the relative differences in mathematical
maturity between the two grades.   The seventh graders,
according to this assumption, are less mathematically mature and hence, they are more attracted to recreational and
reading materials.   The eighth graders, the more mature,

Individualized Instruction in Mathematics

TABLE 1

Percentages Of The Self-Selection Time Spent In Each Activity

	Seventh						Eighth						All	
	7-A		7-B		All Seventh		8-A		8-B		All Eighth		All 7th	All 8th
	F	M	F	M	F	M	F	M	F	M	F	M		
Cultural	35	30	29	28	32	29	27	32	13	24	20	28	30	24
Manual	15	11	0	11	9	11	6	18	30	23	17	21	10	19
Textual	19	15	34	21	25	19	21	10	2	1	13	5	22	9
Conventional	21	34	27	34	25	34	29	30	53	20	40	25	29	33
Explorational	10	10	10	6	10	7	17	10	2	32	10	21	9	15

appear to be more involved in conventional materials most of which require some mathematical knowledge.   The recreational nature and the novelty of most of the cultural activities were intrinsically appealing to many students no matter what their mathematical knowledge.   Hence, we find that eighth graders also spend about 24 per cent of their time on these materials.   Also, the conventional activities included some materials with a remedial nature and hence it might be possible that the seventh graders spent a high percentage --about 29 per cent of their time also on conventional materials for that reason.   The eighth graders' assumed maturity might also be behind the trend that they spent more time, compared to the seventh graders, on advanced units classified under the explorational activities.

b)   Manual activities occupied more of the eighth graders' than it did the seventh graders' time.   Probably the eighth graders could have more patience, skill and fun in constructing a "hide-and-seek" hexaflexagon, stitching some geometrical curve or conducting a baseball survey.   This also might be interpreted as showing an increase in awareness of the eighth graders about the practical needs of devices such as adding machines, geometrical models or surveying method.

c)   Seventh graders spent 22 per cent of their self-selection time on textbook materials.   This is very high when compared with the 9 per cent which the eighth graders spent on their textbook.   This might be because of any beginner's innate respect for and confidence in a textbook.   The seventh graders, who are beginners both in the high school and in a departmentalized system, may be more uncertain of themselves and more eager to conform by studying this "required" textbook which at the same time was the major source for the knowledge on which they were to be evaluated. Another possible reason is that the materials in the SMSG textbooks are in themselves new and more interesting to the seventh than to the eighth graders, who are more familiar with the SMSG textbooks.   It might also be possible that some of the seventh graders used their textbook out of inability to select other materials.

d)   If we consider that the conventional activities were more or less related to either remedial or acceleration work of the students' mathematical level, we can deduce from the high percentage of the time spent on these activities that there is a continuous need and desire on the part of the students to restore the balance of their abilities by either making up their deficiencies or advancing their status to an optimum level. In other words, students in general, whether they use the self-selection method or not, need to have some time to work by themselves in activities which may not present something new but which fill the gap between their own ability as individuals and that which is required from them as members of a certain group. Students in all experimental classes spent about 60 per cent of their self-selection time in textual, conventional, and explorational activities.

STUDENTS' REACTION INVENTORY

Name F
M
Teacher Class

From your experience with self-selection, kindly react to the following items as indicated in each one.

I. Self-Selection Contribution (circle the appropriate answers).
  (A) Self-selection has helped me to
      (1) Work on things which are of real interest to me
      (2) Work to my ability
      (3) Work at my own rate
      (4) Learn without pressure on me
      (5) Feel at home with mathematics
      (6) Have a change from the textbook
      (7) Be able to work independently
      (8) Be able to trace an idea in available references
      (9) Work with other students
      (10) Get more help from the teacher
      (11) Increase my mathematical knowledge
      (12) Broaden my views about mathematics
      (13) Improve my weaknesses in certain areas
      (14) Learn materials for next grade
      (15) Prepare for the regular classwork
  (B) Self-selection has not helped me in any way:
      yes   no
  (C) I feel I was lost in the self-selection periods:
      agree   disagree

    (D)   I feel as if I wasted most of the selection time:
         agree   disagree

II.   Self-Selection Accomplishments (circle one answer only).
My accomplishments in the self-selection periods can
be estimated as (compared to similar time in a regular
classwork)
    (1)   less than the regular classwork
    (2)   approximately as much as the regular class-
         work
    (3)   more than the regular classwork

III. Self-Selection Materials (circle one answer only).
The materials I used in the self-selection periods are
(compared to the textbook)
    (1)   less interesting than the textbook
    (2)   as interesting as the textbook
    (3)   more interesting than the textbook

IV.   General Reaction (circle one answer only).
Have you liked self-selection the way you experienced
it this year?
    (1)   no
    (2)   undecided
    (3)   yes

V.   Please comment here about any aspect of self-selection
or your experience with it

TEACHERS' REACTION INVENTORY

Name   Class

    I would appreciate it if you would kindly react to the
following items concerning the different aspects of self-selection as you experienced it with this class.   Please use the
accompanying five points scales such that:

(1)   For classifications A and B:
    1 = strongly disagree
    2 = disagree
    3 = undecided
    4 = agree
    5 = strongly agree

(2) For classification C (compared to regular class situations):
    1 = very low
    2 = lower
    3 = equal
    4 = higher
    5 = very high
    0 when you cannot judge the case

(A) <u>Procedure</u> in the Self-Selection   <u>1 2 3 4 5</u>
<u>Periods</u>

    (1) Students needed more control and discipline (as compared to regular class situation)

    (2) The efforts necessary by the teacher are more than those necessary for regular class

    (3) Time allotted for self-selection is appropriate (two periods a week)

    (4) Self-selection might function better in higher grades where students are more mature

    (5) Self-selection might function better in lower grades

    (6) It is better not to put any restrictions at all on the students with regard to the continuity and depth of their work.

    (7) It is better that a certain portion of a student's grade be based on his work in the self-selection periods

    (8) Students showed seriousness in their work within the self-selection experience

(B) The <u>Materials Used</u> in the Experiment

    (1) There are different levels of difficulty in these materials

    (2) There is a challenge to the students in these materials

    (3) Students used much of these materials

    (4) Small-size books are appealing to students more than larger size ones

    (5) There must be work sheets to

explain the mathematics   <u>1 2 3 4 5</u>
underlying the mathematical
games

(6) Units prepared by teachers are
more appealing to students (as
compared to the published materials)

(7) There must be more self-checked
materials to facilitate the teacher's job in the "feed-back"
process

(8) Reading ability demonstrated a
crucial problem for better use
of the materials

(9) The materials used were more
appealing to the students than
the textbook

(10) The materials used were in
general appropriate

(C)   <u>General</u>
(1) Self-selection helped the students
to undertake independent work

(2) Students obtained some concrete
knowledge in mathematics from
self-selection

(3) Self-selection helped the students
in their regular classwork

(4) Self-selection broadened the students' views about mathematics

(5) Self-selection helped the students
to "get" acquainted with the
cultural part of mathematics

(6) It helped them to appreciate the
practical use of mathematics

(7) It helped them to improve their
attitudes towards mathematics

(8) In general, the overall value of
the self-selection approach can
be estimated as

(9) My general reaction to the experiment is (1 and 2 = unfavorable, 3 = neutral, 4 and 5 =
favorable)

The teachers, in general, showed favorable reactions towards the experiment as a whole and many of its different aspects, agreeing that self-selection helped the students to undertake independent work, that its materials were appealing to the student, and that they covered different levels of difficulties. They also agreed that the approach requires more effort from the teacher and that there is a need for some refinement in the materials in terms of more explanatory worksheets and self-checking materials.

The teachers all agreed also to base a certain portion of the students' grades on their self-selection work. Although evaluation is important in the education process no matter which method is used, its use in grading the self-selection situation might support the idea associating motivation with grades. The self-selection pattern in learning assumes that it creates intrinsic motivations in the students. However, it seems that extrinsic motivations are also needed.

The findings of this study can be summarized in the following general conclusions:

(1) Self-selection has provided the students with an opportunity in which they proved they could function on a highly individualized basis in their mathematical learning.

(2) Although they were all mathematical, many of the students' activities extended beyond and were of a different caliber than the mathematical concepts and skills usually developed in the conventional class situations. These "extra" mathematical learnings covered a wide range of cultural, manual, and explorational materials which were used for remedial or accelerational work. Different items of these materials were acquired by different individuals in such a way that it did not permit a group test for these self-selected activities.

(3) Since the experimental students did at least as well as the conventional students on the essential and textual concepts and skills, the self-selected learnings which were not covered in these achievement tests were the concrete gain in mathematical growth obtained by the experimental students beyond what the control students obtained.

(4) Students also gained in working independently and in undertaking research-types of activities.

(5) Self-selection, by introducing so many varieties
of activities, could stimulate the many differences in interests and cope to some extent with diversity among students.
Students had a chance to "catch up" on their regular classwork, to "make up" for their weaknesses by doing some remedial work, or to follow up ideas in more advanced materials.

(6) Beside the need for better instruments to measure the many variables involved in the self-selection periods,
the experiment showed that there was a need for better
materials to be prepared for use by individuals.   Brief introductions, or previews, of the existing materials could
have facilitated some of the students' difficulties in choosing
or at least could have saved (were that a virtue) some of
the time spent in "browsing. "

(7) It seemed that the 40 per cent of the time devoted for self-selection periods in the second semester constituted a reasonable portion for most of the students and
teachers.

(8) The students' reactions and estimates of their
work in self-selection did not seem to be dependent on their
general or mathematical abilities.   Differentiation among
their reactions appeared to be associated more with grade
and sex differences, where seventh graders appeared to be
happier, while the most unhappy people came from the
eighth-grade girls' group.

(9) Based on teachers' ratings, the students' self-
selection abilities appeared to be associated with their
mathematical abilities and activities.   General ability did
not seem to be a crucial factor.   The diversity of the self-
selection materials could probably meet the different mental
abilities.

(10) The self-selection periods created an intellectually stimulating environment and provided children with rich
backgrounds in different areas of mathematics which each
student felt that he gained by his own efforts guided by his
own interests and capacities.   It was a learning situation
where the teacher's role was not that of the taskmaster but
that of the guide and counselor.

In general, the self-selection periods attempted to,
and to a great extent succeeded in expanding the student's

insight into mathematics and in himself as a unique organism
learning mathematics.

Notes

1.   A. Binet and V. Henri. "La Psychologie Individuelle."
     L'Année Psychologique 2:411-65, 1895.

2.   P. W. Medawar. The Uniqueness of the Individual.
     London: Methuen, 1957; chapter 8.

3.   Garrett Hardin. "Biology and Individual Differences. "
     In Individualizing Instruction, The Sixty-First
     Yearbook of the National Society for the Study of
     Education, Part 1 (edited by Nelson B. Henry).
     Chicago: University of Chicago Press, 1962; p.
     11-12.

4.   Leona E. Tyler. The Psychology of Human Differences.
     2nd edition; New York: Appleton, 1946; p. 3-14.

5.   S. A. Otis. Otis Quick-Scoring Mental Ability Tests:
     Manual of Directions for Alpha Test. New York:
     Harcourt, 1939.

6.   E. K. Strong, Jr. Vocational Interests of Men and
     Women. Stanford, Calif. : Stanford University
     Press, 1943.

7.   Arnold Gesell. Studies in Child Development. New
     York: Harper, 1948; p. 182-84.

8.   Minnie P. Berson. "Individual Differences Among Pre-
     School Children: Four-Year Olds. " In Individu-
     alizing Instruction, op. cit. , p. 112.

9.   Harold E Jones and Mary Cover Jones. "Individual
     Differences in Early Adolescence. " In Individual-
     izing Instruction, op. cit. , p. 126-144.

10. Lawrence K. Frank. "Four Ways to Look at Poten-
     tialities. " In New Insights and the Curriculum,
     The 1963 Yearbook of the Association for Super-
     vision and Curriculum Development. Washington,
     D. C. : National Education Association, 1963; p.
     24.

11.   Marian E. Breckenridge and Margaret N. Murphy.
      Growth and Development of the Young Child.
      Philadelphia: Saunders, 1963; p. 3-34.

12.   Arnold Gesell.   "The Ontogenesis of Infant Behavior. "
      In Manual of Child Psychology (edited by Leonard
      Carmichael.) New York: Wiley, 1954.

13.   S. A. Courtis.   "Maturation Units for the Measurement
      of Growth. " School and Society 29:683-90, 1929.

14.   Willard Olson.   Child Development.   Boston: Heath,
      1959, p. 198.

15.   Raymond B Cattell.   Description and Measurement of
      Personality. New York: Harcourt, 1946; p. 46.

16.   Dorothy H. Eichorn.   "Biological Correlates of Behav-
      ior. " In Child Psychology, The Sixty-Second
      Yearbook of the National Society for the Study of
      Education, Part 1.   Chicago: University of Chi-
      cago Press, 1963; p. 53.

17.   J. Kagan, H. A. Moss, and I. Sigel.   "The Psycho-
      logical Significance of Styles of Conceptualization. "
      In Basic Cognition Processes in Education. Wash-
      ington, D. C. : Society of Research in Child De-
      velopment, 1962.

18.   Sheldon H. White.   "Learning. " In Child Psychology,
      op. cit. , p. 221.

19.   H. Witmer and R. Kotinsky.   Personality in the Mak-
      ing. New York: Harper, 1953; p. 30.

20.   I. E. Sigel.   "The Needs for Conceptualization on
      Child Development. " Child Development 27:242,
      1956.

21.   R. Stewart Jones and Robert Pingry.   "Individiual Dif-
      ferences. " In Instruction in Arithmetic. Wash-
      ington, D. C. : National Council of Teachers of
      Mathematics, 1960; p. 103-10. [For excerpts
      from this item see Chapter X of this book.]

22.   D. E. Berlyne.   Conflict, Arousal and Curiosity. New
      York: McGraw-Hill, 1960.

23. W. Fowler. "Cognitive Learning in Infancy and Early
         Childhood. " Psychological Bulletin 59:145, 1962.

24. Max G. Wingo. "Important Factors Affecting the Edu-
         cational Program. " Review of Educational Re-
         search 23:123, 1933.

25. Willard Olson. "Seeking, Self-Selection and Pacing in
         the Use of Books by Children. " The Packet p.
         3-10, Spring 1952. (Boston: Heath.)

26. Bernice Ault. The Use of the Self-Selection Principle
         in the Teaching of Arithmetic. Master's thesis.
         Ann Arbor: University of Michigan, 1945.

27. Gladys Crosley and Herbert Fremont. "Individualized
         Algebra. " Mathematics Teacher 55:109-12, 1962.

28. M. J. Brannon. "Individualized Mathematics Study
         Plan. " Mathematics Teacher 55:52-56, 1962.

TWO SELF-SELECTIVE APPROACHES IN
JUNIOR HIGH SCHOOL MATHEMATICS

Henry D. Snyder, Jr.

As research continues in the area of child development, the principle of self-selection is a recurring theme.
According to Shane,

> Self-selection implies the creation of a rich environ-
> ment which is also diversified so as to provide a
> variety of activities or projects from among which
> children can self-select work in which they will en-
> gage (individually and/or in groups) in conjunction
> with a topic or subject which promises to be a sound
> 'center of interest' or 'group interest' compatible with
> the development levels of the group. [1]

Any classroom organization which utilizes this principle must
sacrifice some structure and some of its advantages.

Davis reported a study of an experiment in

self-selection of diet in which 15 institutionalized infants aged six to 11 months were observed for periods of time from one-half to four and one-half years.   These infants were offered a variety of food three times each day.   They were allowed to eat as much or as little as they desired of whatever they chose, and they were not forced to maintain a "normal" diet.   Results showed these children to be at least as healthy as comparable children who had all the advantages of a mother's care.   According to Davis, "The study demonstrated that ... young children [sic] could choose their diets and thrive without adult direction as to just what and how much of these foods they should eat. "[2]

Ryan attempted to ascertain if college students in undergraduate educational psychology classes, when allowed to choose the instructional approach by which they would be taught, would make greater than normal gains in the classroom.   The alternatives were "group participation--student directed, " "independent study--student directed, " and "independent study--teacher directed"; the control classes, also with 54 students, used a combination of the three approaches. The experiment ran for eight weeks and criteria for measuring student success were score on a test over principles covered and ability to apply these principles in a simulated teaching experience.   While there were no differences among the three methods in the experimental classes on either criterion, the free-choice group was significant superior to the control group in both areas.   He concluded that the important variable was not which approach was chosen but that the students were allowed to choose. [3]

Use has been made of the self-selection principle in the teaching of reading, but the results are indefinite.   Anderson, Hughes, and Dixon compared a self-selection-based individualized reading program with a basal reader approach and observed rates of growth over a long period of time. The basal reader group of 434 students learned to read earlier than did the 211 students in the self-selection classes and maintained this superiority for some time; however, at mean age 132 months the experimental students were reading at the same level as were their control class counterparts. There was some evidence that the students from the individualized program had a more positive attitude toward reading than did the students in the control classes. [4]   Other studies have shown that in the long run there is little difference between the two approaches in the area of reading.

Ault used a self-selection-based program in teaching third- and fourth-grade arithmetic and compared it with the results from preceding years. The students were allowed to inspect a wide age range of workbook materials and to select the ones they felt were most appropriate. She found that the choices made by the students were appropriate to their ability levels, that the gains in achievement were greater than they had been in previous years, and that the students seemed to have good attitudes toward the program itself. [5]

Ebeid [see previous article] studied the behavior and the choices made by seventh- and eighth-grade students in classes which allowed part-time self-selection of mathematics materials. Seventy per cent of the class time was spent in organized study of School Mathematics Study Group (SMSG) materials, and 30 per cent of the time (two days each two weeks during first semester, two days each week during second semester) was spent studying independently from a variety of materials. In this collection were commercial enrichment materials, text materials, mathematically-oriented non-textbook materials, puzzles, and games.

During the 30 per cent of the class time which was allocated for self-selection the students were expected to pursue some mathematical topic of interest to them. While many students used this opportunity to study topics such as algebra and topology, a few of them spent the time studying from their own SMSG textbook.

A comparison of the experimental classes with the control classes studying the SMSG text full-time showed no significant differences on the SMSG or Sequential Test of Educational Progress--Mathematics (STEP) tests although the experimental classes tended to attain slightly higher scores on the STEP test. In addition, while the attitudes toward mathematics of the control classes tended to decline during the year, those of the students in the experimental classes did not. These results indicate that a 30 per cent reduction in organized classroom time had no adverse effect on achievement; it is probable that they learned much during the self-selection periods that neither the SMSG nor the STEP test measured. It appears also that when students were allowed to choose part of their mathematics programs and class time was used in this fashion, attitudes toward mathematics did not deteriorate as they did in the more highly structured control classes.

Considering the results within the experimental classes, some grade and sex differences became apparent. Only 12 per cent of the students were unfavorable to the part-time self-selection approach; these consisted mainly of eighth graders with eighth-grade girls the least favorable. He found also that the seventh graders were more able to handle the self-selection periods in a constructive fashion than were the eighth graders, although there were a few students who seemed unable to cope with such an unstructured situation. [6]

Fitzgerald used this approach in grades seven and eight in a full-time commitment to the self-selection principle. No conventional classroom instruction was provided during the year; the students were expected to select from a wide variety of materials those topics which were of most interest to them. The material used by Ebeid the preceding year was supplemented by a wider range and variety of material. Among the additions were textbooks from grade level four to college freshman level, commercial programmed materials, SMSG textbooks separated into single-chapter units, commercial workbooks, reading books concerned with mathematics, and short supplementary units written by the teachers of the experimental classes. The classes met four periods per week all year in this fashion; this was the full-time program and all students were included.

Comparisons were made between the experiment classes and the control classes studying SMSG materials in a conventional five day per week program. Final achievement results on the Snader General Mathematics Test showed that seventh-grade boys and girls in control classes were superior to those in the experimental classes in this respect; no such differences were found for the eighth-grade students. When the results were considered by high and low intelligence levels (i. e. , above or below 115 IQ), the brighter boys and girls in the control classes were superior to their experimental counterparts in achievement; no such differences emerged for students of lower intelligence.

Results from the attitude tests showed that the experimental seventh graders demonstrated higher attitudes than did the control seventh-grade students; there were no significant differences among eighth-grade students. Lower intelligence girls in experimental classes maintained higher attitudes than did those in control classes, but there were no differences among boys or higher intelligence students in general.

It appears, then, that the full-time self-selection program restricted achievement for seventh grade and higher intelligence students, but led to not significantly different results for eighth graders or for students of lower intelligence. The program, however, seemed to have a positive effect upon attitudes of seventh graders and of eighth-grade girls of lower intelligence while it produced no different results for the other samples.

Much has been learned from these self-selection from these self-selection studies in the area of mathematics, but many questions have been raised and left unanswered. Which students benefit from which self-selection programs? Are different amounts of time devoted to self-selection more appropriate for different students? Should students have the responsibility of determining whether or not they will be involved in a self-selection program? These questions and others form the rationale for the current study.

Experimental Treatments

The studies by Ebeid and Fitzgerald in the area of junior high school self-selection of mathematics programs showed their approaches to be promising for many students but quite ineffective for many others. Since neither of the prior programs offered the students a teacher-taught option during the self-selection periods, this option was provided in the current study to determine if such an arrangement would meet the needs of more of the students.

Treatment E_1. The first experimental treatment (E_1) was a modified form of Fitzgerald's self-selection program. Within certain flexible guidelines the students were allowed to choose the mathematics programs they would pursue; these guidelines formed the pacing part of the program.

One of the choices available to the students was a conventional class setting in which they would follow a textbook in connection with a teacher-taught program. The lessons were planned so that the textbook would be completed during the year and in such a way that the teacher's presentation would cover from 25 to 35 minutes every other day. The remainder of the time was allowed for directed study, thus permitting the teacher to assist these students as well as the students who worked independently of this class setting.

Assignments were prepared every other day and placed on ditto sheets for class distribution.   Three levels of assignments were prepared and all were placed on the assignment sheet.   Each student who had chosen the teacher-taught option was expected to select one of the three assignment levels for himself.

These differentiated assignments, called Level 1, Level 2, and Level 3, were formed on the bases of depth and difficulty rather than length.   The most demanding assignments, Level 3, were prepared for the more able and highly motivated students.   While some basic content and skill work was included, the greatest emphasis was in the direction of extension of the current topics, anticipation of future work, and application of these topics in other contexts.

The assignments of Level 1 were formulated with the less able, less motivated students in mind.   Understanding of the basic content was the objective, so the assignments of this level stressed redevelopment and review of previous material and skill work in connection with the current topic; as much as possible these exercises were placed in novel settings to enhance interest and de-emphasize the review nature of the content.

Level 2 assignments were intended to be a middle ground between the other two.   They usually contained fewer skill-oriented problems than did Level 1, more development of the current class topic than did either of the other levels, and less extension of the subject matter than did Level 3 assignments.

Students were free to select any assignment level they wished.   Although the above descriptions indicate for which students the assignments were prepared, no pressure was exerted upon students to select any particular level.   The students were encouraged to examine the problems from each level and to choose the one which "seemed to be the best one" for themselves.

As assignments were completed they were checked by the students from answer sheets posted on a bulletin board in the classroom, were turned in to the teacher who scanned

each one and recorded its being completed, and were returned to the students who placed them in inidividual file
folders.

     The other basic choice available to the students in
the E_1 treatment classes was a self-selection-pacing program in which each student could, in consultation with the
teacher, select, structure, and pursue a mathematics program of his own.   Source material for such independent work
was available in the large bookcase in the room.   The students were not restricted to these materials; any other
sources for ideas were admissible.

     In general, a student pursuing this option proceeded
in the following manner.   He found an area of interest from
a prior interest, from his reading, or from suggestions of
other students.   (Early in the year when a student claimed
he could find nothing of interest, the teacher would offer a
suggestion; after approximately two months the teacher's
suggestion was, "How about working with the class for a
while?").   Once the student had identified an interest he
came to the teacher with the idea and together they arrived
at an appropriate mathematical area and identified the skills
that would be necessary in such a study.   Then the student
looked for materials available on that topic and, in consultation with the teacher, set up a tentative time schedule for
his study of this topic.   Finally he began his study of the
topic.

     There were flexible provisions made for mobility be-
tween the two options of the E_1 treatment.   Students were
urged to carry their topic of study to completion, but were
encouraged to consider both options when a new topic was
to be begun.   They were urged to reevaluate periodically
their needs and interests and to choose the program which
seemed more appropriate.   This reevaluation and flexibility
feature formed the cornerstone of the pacing program in the
E_1 treatment classes, although the specific pacing procedures
were slightly different for the two options.

     For the students working independently the pacing
procedure began with the method of choosing a study topic
and setting a time schedule.   A weekly progress report was
required of each of these students comparing actual progress
with their schedule.   By observing these students in class,
studying their progress reports, and looking at completed
work in their folders the teacher was able to keep track of

their progress and to provide help, encouragement, and advice when it was needed. Periodic scheduled conferences
with each of these students provided an opportunity for looking at long-range plans and for looking back to evaluate
progress. These students were kept informed of the topic
being covered in the conventional class, of what other students were studying, and of other possibilities of topics related to their own.

Treatment E_2. The second experimental treatment
(E_2) was a modified form of a conventional classroom setting. The basic textbook for the course was the SMSG
Mathematics for Junior High School and everyone was responsible for this material.

The same assignment sheets were used as in the E_1,
treatment classes and each student was to select the assignment level which seemed most appropriate for him. Again,
no pressure was exerted on any student to select any particular level of assignment. As in the E_1 treatment, the students corrected their own work, the teacher looked at and
recorded it, and the students placed it in their own file
folders. The pacing procedures were the same as those for
the E_1 treatment students who selected the teacher-taught
option.

The "choice" aspect, the characteristic of this program which allowed the student some control over his own
mathematics program, was two-fold. First, as in the E_1
treatment, the student selected his own assignment level.
The most important characteristic of the E_2 approach, gave
the student the opportunity to enrich the class program with
a part-time independent study of another topic; in fact, the
students were encouraged to do this. They had the option,
then, of substituting the study of a topic interesting to them
for some of the depth of the classroom topic. In such a
case this could be accomplished either by using home-study
or by choosing a less-demanding assignment level to allow
more time for independent study. The pacing procedure for
students who undertook this option was the same as for students in the independent study option of the E_1 program.

While it was possible (and did happen in several
cases) that students in the E_1 and E_2 classes would pursue
identical mathematics programs for a length of time, the
treatments had different emphases. Basically the E_1 treatment was one in which a student could determine his own

mathematics program and pursue it in depth or with breadth. The E_2 treatment was one of a prescribed program of study which the student could supplement as he wished.   In both programs approximately 75 per cent of the class time was devoted to independent study.   Teacher presentation occupied one-half hour each two days; during the remaining independent study time the teacher was available for individual help for all students.

Objectives

This study compared two experimental junior high school mathematics programs in which students were given a certain degree of freedom in and responsibility for their own mathematics program.   In one program the emphasis was on a self-selection--pacing approach with the emphasis on independent work; in the other the emphasis was on a self-selection--pacing approach stressing enrichment.

Specifically, the objectives were to

1.   Attempt to identify the characteristics (e. g. , mathematical achievement, attitude toward mathematics, dependence-proneness, mental age, chronological age, arithmetic age, IQ, sex, grade level, public versus private school background, personality) of students who engage in individual self-selected learning activities in junior high school mathematics when given such an alternative, and of those who do not.
2.   Study the effects of the two self-selection--pacing programs upon certain measurable characteristics (e. g. , mathematical achievement, attitude toward mathematics, dependence-proneness) of participating students.
3.   Compare changes in the achievement, attitude, and dependence variables for the students in the two self-selection--pacing programs with those of students in:
a.   the other experimental treatment
b.   Fitzgerald's experimental classes (full-time self-selection)
c.   Ebeid's experimental classes (part-time self-selection)
d.   public school control classes.
4.   Attempt to determine the effectiveness of differentiated assignments in such settings as these.
5.   Determine student reaction to the self-selection--pacing approach in junior high school mathematics, and compare this with reactions of students to previous self-selection

programs.

Treatments E_1 and E_2

Tables 1 and 2 show means $(\overline{X})$ and standard deviations (s) of the per cent of time spent by students from various subgroups working on topics independently for the classes under Treatment E_1 and E_2.

Summary Of The Findings

Data for analysis were gathered throughout the study. As part of the school's regular testing program the California Test of Mental Maturity was administered to all students during the first week of school in the fall.   Pretests of mathematical achievement, Sequential Test of Educational Progress--Mathematics and the Snader General Mathematics Test, of attitude toward mathematics, Opinionnaire, and of dependence-proneness, It's A Matter of Opinion, were given during the first month and posttest scores were obtained during the final month of school in the spring; the SMSG mathematical achievement test, SMSG Mathematics Inventory, was administered at the end of the year only.   Records of student work throughout the year provided additional data.

Following are summaries of the findings from data concerning each of the null hypotheses.

H_1:   There is no identifiable pattern of characteristics to predict which students will engage in self-selected mathematics learning activities and those who will not for either treatment E_1 or treatment E_2.

Through a linear stepwise regression analysis an attempt was made to find which of the data available at the beginning of the year could be used to predict which students would actually participate in the independent study option of each program.   For each of the subgroups a prediction equation and a set of correlation coefficients were obtained with regard to the variable "per cent of time spent working on independent study topics."   The findings were:

(1)   Boys sent significantly more time studying independently than did girls in E_1 treatment sections $(p < .01)$.
(2)   Predictor variables chosen for the dependent

TABLE 1

Treatment E_1: Per Cent Of Time Spent
Working On Independent Study Topics

Group	n	$\overline{X}$	s
Total	48	49.87	38.42
Seventh Grade	23	47.55	33.22
Eighth Grade	25	52.00	43.23
Boys	24	62.92	34.54

TABLE 2

Treatment E_2: Per Cent Of Time Spent
Working On Independent Study Topics

Group	n	$\overline{X}$	s
Total	50	9.20	19.20
Seventh Grade	25	10.20	23.07
Eighth Grade	25	8.20	14.78
Boys	25	9.00	20.56
Girls	25	9.40	18.16

variable in three or more of the prediction equations among E_1 treatment subgroups were reading age, stability of personality, reasoning factor of arithmetic age, spelling age, Snader pretest score (emphasizing computational skills), and language factor of mental age.

(3) Significant correlation coefficients relating independent variables with the dependent variable for the E_1 treatment group were arithmetic age and its reasoning factor, Snader pretest score, STEP pretest score (emphasizing reasoning abilities), mental age and its language factor, and stability of personality.

(4) Predictor variables chosen for three or more

prediction equations for E_2 treatment students were the fundamentals factor of arithmetic age, Snader pretest score, nonlanguage factor of mental age, and stability of personality.

(5) Significant correlation coefficients relating independent variables with the dependent variable for the E_2 treatment group were Snader pretest score, reasoning factor of arithmetic age, spelling age, and STEP pretest score.

On these bases the null hypothesis is rejected and it is concluded that certain characteristics measured by the independent variables provide an indication of whether or not a student will pursue independent study in junior high school mathematics when given such an opportunity.

$\underline{H_2}$: The experimental treatments E_1 and E_2 will not have different effects upon the mathematical achievement, attitude toward mathematics, or dependence-proneness of any particular student.

Pairs of students matched on sex, mental age, arithmetic age, and grade level were identified and one member of each pair was placed in each treatment section.   t-tests on the gain scores from the Snader and STEP achievement tests, the attitude toward mathematics scale, and the dependence-proneness scale were obtained; also a t-test on the SMSG posttest scores was calculated.   The findings were:

(6) No significant differences were found between Snader test gain scores for the two treatments.   With respect to this measure of mathematical achievement the null hypothesis is not rejected.

(7) Significant differences in gains on the STEP test were found only for lower intelligence girls and lower intelligence seventh-grade students.   For these two samples the null hypothesis is rejected and it is concluded that for girls and seventh-grade students of lower than average intelligence the E_1 treatment led to greater gains in STEP test scores than did the E_2 treatment.

(8) None of the differences in SMSG test scores were statistically significant.   The null hypothesis for this measure of mathematical achievement is not rejected.

(9) Changes in attitude toward mathematics were more positive in the E_1 treatment for eighth-grade girls

(particularly those of higher intelligence) than in the E_2 treatment, but the E_2 treatment led to more positive changes for seventh-grade girls of lower intelligence. For these segments of the experimental population the null hypothesis dealing with attitudes toward mathematics is rejected and it is concluded that the E_1 treatment led to more positive changes in attitude than did the E_2 treatment for eighth-grade girls ($p < .05$), while the E_2 treatment led to more positive changes in attitude for lower intelligence seventh-grade girls ($p < .05$).

(10) Changes in dependence-proneness scores indicated that the E_1 treatment led to greater gains in dependence than did the E_2 treatment for seventh-grade students (particularly those of higher intelligence). Eighth-grade boys (particularly those of higher intelligence) became more dependent in E_2 treatment classes.

The null hypothesis pertaining to dependence-proneness is rejected for seventh-grade students and for eighth-grade boys and it is concluded that seventh-grade students became more dependent in E_1 treatment classes than in E_2 treatment classes ($p < .05$) while eighth-grade boys became more dependent in E_2 treatment classes than in E_1 treatment classes ($p < .05$).

$\underline{H}_3$:   There will be no significant differences in the changes in mathematical achievement, attitude toward mathematics, or dependence-proneness between either experimental treatment and:
  a.   the other experimental treatment
  b.   Fitzgerald's experimental classes (full-time self-selection)
  c.   Ebeid's experimental classes (part-time self-selection)
  d.   public school control classes.

Pretest-posttest gain scores on the Snader and STEP tests of mathematical achievement, the attitude toward mathematics scale, and the dependence-proneness scale (and the end-of-year scores on the SMSG achievement test) were compared through analyses of variance and covariance techniques to determine if either experimental treatment led to significantly different results than did the other. Results were compared with those of Ebeid and Fitzgerald whenever data were comparable to determine if significantly different results were obtained; in the latter comparisons the t-test was used.

(11)   Gain scores on the Snader test were not significantly different for corresponding sections of the E_1 and E_2
treatment groups.   The null hypothesis is not rejected for
this measure of mathematical achievement.

(12)   Significant differences between STEP test gain
scores showed that for combined grade level groups and for
students of lower intelligence the E_1 treatment led to greater
gains than did the E_2 treatment; no other significant differences were found for STEP test gains.   The null hypothesis
is rejected for the total group and for the lower intelligence
sample and it is concluded that total groups and lower intelligence students achieved greater STEP test gains in E_1
treatment classes than in E_2 treatment classes ($p < .05$) in
both cases.

(13)   No significant differences arose between the
SMSG test scores for the two treatments.   The null hypothesis is not rejected for this measure of mathematical achievement.

(14)   Gain scores for attitude toward mathematics
showed that eighth-grade girls' attitude scores decreased
significantly more in the E_2 class than in the E_1 treatment
class.   The null hypothesis concerning attitudes toward
mathematics of eighth-grade girls is rejected and it is concluded that changes in attitude for eighth-grade girls were
less negative in E_1 treatment classes than in E_2 treatment
classes ($p < .05$).

(15)   Changes in dependence-proneness scores showed
that students of lower intelligence (particularly in eighth-
grade sections) became more dependent in E_1 treatment
classes while they became more independent in E_2 treatment
classes.   With respect to the dependence-proneness of students of lower than class average intelligence the null hypothesis is rejected and it is concluded that these students
became more dependent in E_1 treatment classes while they
became more independent in E_2 treatment classes (the resulting difference was significant ($p < .05$).)

(16)   There were no significant differences between
STEP pretest scores of either of Ebeid's groups and those
of the current experiment; significant differences on posttest
scores showed that

    a.   the E_1 treatment led to higher scores than did either Ebeid's experimental or control treatments for total groups

    b.   the E_1 treatment led to higher scores at seventh-grade level than did Ebeid's control treatment.

The null hypothesis is rejected for this measure of mathematical achievement and it is concluded that the E_1 treatment led to higher STEP test scores than did either the experimental or control treatments reported by Ebeid ($p < .05$ and $p < .01$, respectively).

(17)   End-of-year SMSG test scores were significantly higher for Ebeid's experimental class seventh-grade students than were corresponding scores for E_1 treatment seventh graders. The null hypothesis is rejected for this measure of achievement of seventh-grade students and it is concluded that these students scored higher on the SMSG test after experiencing Ebeid's experimental treatment than after experiencing the E_1 treatment ($p < .01$).

(18)   Pretest scores of attitude toward mathematics were significantly higher for E_2 treatment groups than for either Ebeid's experimental seventh grade or combined grade level groups; no other significant differences were found among attitude pretest scores. Posttest scores of attitude toward mathematics showed that

    a.   the difference disappeared for combined grade level groups

    b.   the difference was still significant for seventh graders

    c.   the final attitude scores of E_2 treatment seventh-grade students were significantly higher than those of Ebeid's control seventh-grade classes.

The null hypothesis concerning attitude toward mathematics is rejected for comparisons concerning Ebeid's part-time self-selection study and it is concluded that attitude changes were more positive for Ebeid's total experimental group than for the E_2 total group, but these changes were more positive for the seventh-grade students in the E_2 class than for seventh graders in Ebeid's control classes.

(19)   Gain scores on the Snader test show that Fitzgerald's control classes achieved significantly greater gains than did either the E_1 or E_2 treatment classes for total groups, for seventh grade students, for boys, and for higher intelligence samples. The null hypothesis concerning this measure of achievement is rejected and it is concluded that

Fitzgerald's control treatment led to greater Snader test gains for total groups, for seventh-grade students, for boys, and for higher intelligence students than did either E_1 or E_2 treatments ($p < .01$ in all cases).

(20)   Gain scores on the Snader test showed the E_1 treatment led to significantly greater gains than did Fitzgerald's experimental treatment for girls and for eighth-grade students; among eighth-grade girls gain scores were higher for the E_2 treatment than for Fitzgerald's experimental treatment.   The null hypothesis for this measure of mathematical achievement is rejected and it is concluded that gains on the Snader test were lower for Fitzgerald's full-time self-selection program than for the E_1 treatment among eighth-grade students and among girls, and lower than for the E_2 treatment among eighth-grade girls ($p < .05$ in all cases).

(21)   STEP test scores were significantly higher for the E_1 experimental classes than for Fitzgerald's experimental treatment among eighth-grade students, boys, and both high and low intelligence samples.   The null hypothesis is rejected for this measure of mathematical achievement and it is concluded that the E_1 treatment led to higher STEP test scores than did Fitzgerald's full-time self-selection treatment for eighth-grade students, for boys, for higher and for lower intelligence students ($p < .05$ in all cases).

(22)   Attitude gain scores showed that Fitzgerald's experimental treatment led to greater gains than did the E_2 treatment among girls.   The null hypothesis is rejected with respect to the attitude changes of girls and it is concluded that greater gains in attitude toward mathematics were made by girls in Fitzgerald's experimental classes than by those in E_2 classes ($p < .05$).

(23)   Gains in dependence proneness indicated that the E_1 treatment led to greater dependence while Fitzgerald's control treatment led to greater independence for combined groups, for seventh graders, and for lower intelligence students.

The null hypothesis is rejected and it is concluded that the E_1 treatment led to significantly greater gains in the direction of increased dependence-proneness than did Fitzgerald's control treatment among total groups ($p < .05$), among seventh-grade students ($p < .01$), and among lower intelligence students (p < . 01).

$\underline{H}_4$:   When correlation coefficients relating arithmetic age
        and mean assignment level chosen are compared for
        subgroups identified by experimental treatment, grade
        level, sex, or intelligence level,
        a.   the coefficients for any subgroup for the two
             semesters will not be different, and
        b.   the coefficients for the various subgroups will
             not be different for either semester.

For each student in the E_2 treatment sections and
for those E_1 treatment students who selected textbook assignments during a large portion of both semesters, a "mean
assignment level chosen" was determined for each semester
and these were related to their arithmetic ages by means
of product moment correlation coefficients.   The coefficients
obtained for each semester were compared to determine if
significant changes in this relationship occurred for any
particular group of students.   The findings were:

(24)   Seventh-grade students in the E_1 treatment class
reached a greater correlation between arithmetic age and
mean assignment level chosen during second semester than
during first semester.

(25)   For E_2 treatment students there was a significant correlation between arithmetic age and mean assignment
level chosen during both semesters; for E_1 treatment students these correlations were significant only for eighth
graders and for students of lower intelligence.

(26)   There was a significant correlation between
mean assignment levels chosen during the two semesters
for students in both E_1 and E_2 treatment sections.

(27)   Seventh-grade students in E_2 treatment classes
achieved a significantly higher correlation between arithmetic
age and mean assignment level chosen during the first semester than did seventh graders in E_1 treatment classes.

(28)   In E_1 treatment classes eighth-grade students
achieved a significantly higher correlation between arithmetic
age and mean assignment level chosen than did seventh graders during first semester.

(29)   In E_2 treatment classes the correlation between

arithmetic age and mean assignment level chosen was significantly higher for students of higher intelligence than for
those of lower intelligence during both semesters.

The null hypothesis is rejected and it is concluded
that the correlation between arithmetic age and mean assignment level chosen was greater in E_2 and in E_1 treatment
classes, was greater for eighth-grade students than for
seventh graders in E_1 treatment classes, and was greater
for higher intelligence than for lower intelligence students in
E_2 treatment classes.

$\underline{H_5}$:   In response to the question, "On the whole, what did
you think of the mathematics program this year?" there
will be no significant differences in the proportions of
students having positive, positive with reservations, or
negative reactions for either of the experimental treatments or for Fitzgerald's or Ebeid's experimental
classes.

At the end of the year in each of the three studies,
the students were asked to give their reactions to the programs.   In Fitzgerald's full-time self-selection program the
question asked was, "On the whole, what do you think of
self-selection in mathematics?"   In his part-time self-selection experiment Ebeid asked "Have you liked self-selection
the way you experienced it this year (1) no, (2) undecided,
or (3) yes?"   Responses in all three cases were classified
into three categories which are assumed to be equivalent
to "positive," "positive with reservations," and "negative."
By means of a chi-square analysis the numbers of students
responding in the various categories were compared to see
if there was reason to believe that the different programs
produce different reactions as measured by this question.
The findings were:

(30)   When the four experimental treatments were
considered together (E_1, E_2, part-time and full-time), the
resulting distributions of responses were significantly non-
homogeneous for total groups, eighth-grade students, boys,
and girls.   The aspect of the null hypothesis which concerns
all four studies is rejected and it is concluded that the different treatments did not lead to the same proportion of
positive responses for total groups ($p < .01$), eighth grade students ($p < .01$), boys ($p < .01$), or girls ($p < .05$).

(31)   There was no significant difference in proportions

of positive responses between the E_1 and E_2 treatments for any of the samples. The aspect of the null hypothesis relating results of the two current experimental treatments is not rejected.

(32) The E_1 and E_2 treatments led to a higher proportion of positive responses than did the full-time self-selection treatment for eighth-grade students ($p < .05$).

(33) The E_1 and E_2 treatments led to a lower proportion of positive responses than did the part-time self-selection program for total groups ($p < .01$), for eighth-grade students ($p < .05$), and for boys ($p < .05$).

Notes

1. H. Shane. "Grouping in the Elementary School." Phi Delta Kappan 41:313-19, 1960.

2. C. M. Davis. "The Self Selection of Diet Experiment-- Its Significance for Feeding in the Home." Ohio State Medical Journal 34:862-68, 1938.

3. T. A. Ryan. "Testing Instructional Approaches for Increased Learning." Phi Delta Kappan 46:534-36, 1965.

4. I. H. Anderson, B. O. Hughes, and W. R. Dixon. "The Relationship Between Reading Achievement and the Method of Teaching Reading." University of Michigan School of Education Bulletin, Number 27, 1956.

5. B. Ault. The Use of the Self-Selection Principle in the Teaching of Arithmetic. Master's thesis. Ann Arbor, Mich.: University of Michigan, 1945.

6. W. T. Ebeid. An Experimental Study of the Scheduled Classroom Use of Student Self-Selected Materials in Teaching Junior High School Mathematics. Doctoral dissertation. Ann Arbor, Mich.: University of Michigan, 1964. [See article immediately preceding for excerpts from this work.]

7. W. M. Fitzgerald. Self-Selected Mathematics Learning Activities. U. S. Office of Education Cooperative

Research Project Number 2047.   Ann Arbor, Mich. : University of Michigan, 1965.

In this chapter three additional examples of reports of research into the merits of individualization are presented for examination and study. Not all the results reported are favorable, and it is important to realize that not all research on something as ephemeral as a method of instruction will result in identical findings. I do not regard a certain proportion of negative findings as either surprising or alarming. There are present in every research study concerned with styles of teaching or of learning innumerable confounding factors which may very well lead to seemingly contradictory findings. Whatever claims may or may not have been made on behalf of individualized instruction, a claim that this approach represents a cureall or panacea for all educational ills or for every educationally troublesome area is not among those advanced in this anthology or by any person genuinely informed about the nature of such instruction.

Richard M. Bradley

The first portion of this chapter is about one of the most common practices in American schools. One can compile a long bibliography of expository discussions about the merits and demerits of various plans for assigning homework, but the thesis from which this excerpt is taken is one of only a half dozen or so that report research studies on this almost universal American school practice.

Professor Bradley, of Clarion State College in Pennsylvania, acquired his enthusiasm for and belief in individualization of instruction while teaching grades one through eight in an Amish one-room school in Lancaster County, Pennsylvania. Later one he spent several years first as an elementary school supervisor and later as director of elementary education in that county. After obtaining his doctorate at Temple University in 1967, he moved to his present position.

It has always seemed to me that it would make sense to think about individualizing homework assignments, and I have often spoken about this in both in-service and pre-service classes.   One can certainly make what seems to me to be a very sound theoretical argument in favor of this concept.   It is quite interesting, therefore, to find a study that specifically investigates this very point.   A number of the findings of this study reveal a statistically significant difference in certain factors which are favorable to the use of individualized homework assignments, but it must be noted that the findings also show that this was far from a universal finding; in some of the experimental schools it was not at all true.   In addition, there were certain factors of interest and the like where one might well expect that individualization would have shown marked effects but where the data shows that such effects did not in fact exist.   It is possible that the individualization of homework assignments was not carried out as skillfully in some schools as in others.   One may also wish to consider that this procedure was an innovative one for both teachers and pupils.   Nevertheless, the possibility that the benefits of this procedure are not as great as one might have supposed cannot be completely set aside.

Cecil T. Nabors

The next passage in this chapter was written by Professor Nabors of Sam Houston University in Huntsville, Texas.   It is taken from his doctoral dissertation completed at the University of Houston in 1968.   Dr. Nabors teaches courses in mathematics education and supervises student teachers in the Education Department at Sam Houston University.   He brings to his present duties a broad background of teaching mathematics at the elementary, secondary, and college levels, and was an elementary school principal for 12 years.

Anyone who has been a teacher in the elementary school will, I think, agree that one of the phases of arithmetic that gives a disproportionate amount of trouble to many youngsters is the solution of verbal arithmetic problems.   One confounding factor in examining possible reasons for this is that both reading and mathematics competencies are involved.   The study by Dr. Nabors reports on an

experiment using 316 fifth graders in five elementary schools
as subjects in which a comparison was made of the relative
effectiveness of the use of individualized verbal problem assignments and the use of regular fifth-grade textbook material. It is interesting to note that the greater gains made
by the individualized groups were confined specifically to
problem-solving and did not extend to performance on a
generalized test on arithmetic concepts. This test demonstrated no significant differences between the scores made
by the experimental and control groups.

Maria Paul Morrison

        The last item in this chapter is by Maria Paul Mor-
rison who, at the time that she wrote the article excerpted
here from Education, in 1937, was a teacher at the Hosmer
School in Watertown, Massachusetts. Despite diligent efforts to locate the author, I have been unable to do so and
therefore cannot include any further specific information
about her. It seemed to me, however, that readers would
find this article, which was published over three decades
ago, of more than passing interest.

        One of my professors at Columbia University, when
I was working on my doctorate, was Professor Paul Mort,
who had determined that the time required for an educational
"invention" or innovation to become generally accepted was
close to 40 years, and often even longer. This article
serves to emphasize the truth of his statement, for the concept described in this article is almost identical to the concept we are discussing in this volume so many years later.

INDIVIDUALIZED VERSUS UNIFORM HOMEWORK

Richard M. Bradley

        How does progress in mathematics achievement for
students assigned individualized homework compare with
progress of students who receive blanket-type assignments?

        In consideration of the immeasurable hours of human
effort engaged in this practice, by both pupils and teachers,

it is imperative that homework be assigned in only the most
judicious manner. In light of the nature of the practice, it
is not surprising to find a variety of opinions as to just what
constitutes a judicious manner. What does seem somewhat
incredible, is the inadequate attention given the matter by
educational research.

The problem has been complicated further by the
public mood of the post-Sputnik era. The general confusion
which equates quantity with quality in education has resulted
in an almost intolerable situation with respect to homework.
The size and frequency of assignments have too often been
in excess of good judgment and blame is frequently attributed
to pressures exerted by parents. However understandable
or valid this reasoning may be, refusal by educators to accept this responsibility is unprofessional. When the failure
is in respect to the quality and type of homework, the matter is even more serious. Responsibility for this latter
shortcoming, to the extent that it exists, must rest solely
with the professional and there can be no scapegoat.

The practice of dispensing blanket-type assignments
to all members of a given class would seem to leave the
matter of quality somewhat open to question. The mathematics program in the elementary school, for example, typically follows this pattern. Each child in a class is expected
to do the 24 problems on a given page regardless of whether
those 24 problems may present him with an exercise in futil-
ity or a confrontation with sheer boredom. Practices of
this nature appear to be in direct conflict with what is presently known about the teaching-learning process. Nevertheless, such practices persist. As early as 1949, the highly
regarded Pennsylvania Bulletin 233-B suggested that "assignments should be individual since the speed of work (of pupils differs. "[1] Strang has for some time stressed the need
for individualizing assignments. [2] Unfortunately, in actual
practice, the individualization of homework has gained only
an attenuated usage, and that is limited to situations where
individual considerations are not completely ignored.

Basic to the hypothesis of this investigation is the assumption that failure to personalize homework is the real
source of a considerable number of problems frequently attributed to other causality. Abolishing homework for all
children does not appear to be a solution to the real problem.

Unfortunately, no scientific research is available regarding the individualizing of homework. In fact, very few
aspects of the problem have been adequately treated, as at-
tested to by pleas from practically all who have inquired into the matter. Although hundreds of articles have been published concerning the topic, very few provide much more
than the opinion of the author. Goldstein examined each of
the 280 articles on homework listed in the Education Index
for the 30-year period before 1958 and reported that only
17 articles proved to be original reports of experimental
research. [3] As a preparation for the present investigation,
this researcher continued the investigation, updating the
same to include all material listed in the Education Index
until December, 1966. From the total listing of 362 articles, 19 were reports of experimental projects. Most of
these studies were concerned primarily with the amount of
homework assigned and none examined the matter in terms
of individualized assignments.

There is an unseemly shortage of scientific research
evidence on homework and, also, reason to question the
validity of the limited amount which does exist. By contrast, and possibly due in part to this shortage, there is an
abundance of literature reflecting a subjective point of view
on this controversial educational practice. A considerable
amount of what has been written is devoted to the timeworn
pro and con debate, a controversy which is not limited to
the United States. Those who oppose the practice believe
it is unproductive and suggest that it may even be harmful
to the child's health or well-being. Those desiring to see
the practice increased view it as a necessary discipline and
in some respects an educational panacea.

Most of the information obtained from survey type
studies is concerned with the attitude of parents, teachers,
and pupils toward the subject. In general, they report that
parents want and approve homework although parents believe
homework should be geared to the child's ability and should
take family living into account. One study found that teachers at the secondary level placed more emphasis upon homework than did elementary teachers in determining grades.
The majority of students reporting in the surveys examined
indicated that homework practices were satisfactory although
junior high school pupils complained about too many assignments. Nineteen experimental studies have been reported
in the educational literature, the majority of these were
conducted during the 1930's. The ten studies of that period

were almost entirely devoted to the elementary school level
and all but two reported results generally unfavorable to
homework.   During the 1940's and 1950's research activity
on the topic decreased markedly as did the subjective literature.   The five studies during those two decades were directed toward the problem at the secondary school level and
three of the five investigations reported results in favor of
the practice.   In the present decade, four experimental
studies have been reported, three of which were primarily
concerned with the amount of homework assigned.   The
three were probably initiated as a response to the post-
Sputnik "pile-it-on" policy.   The fourth study was concerned
with the nature of homework assigned, an area in critical
need of attention.   Unfortunately the results of the fourth
study were inconclusive and the other three produced conflicting evidence similar to the situation in earlier years.
On balance, results of experimentation on homework have
been unfavorable to the practice, although a number of
studies at the secondary level have reached conclusions favorable to the practice.

Definition of Terms

        HOMEWORK: work which the pupil is assigned and
required to do on his own time as an extension of his
classroom work.

        INDIVIDUALIZED HOMEWORK: assignments related
to the work of the classroom but based primarily on the
needs of the individual child with full recognition that the
difficulty, amount, and nature of the work may at times be
similar to and at other times quite different from that of
other members of the class.

        BLANKET-TYPE HOMEWORK: a term used to de-
scribe the practice of assigning homework to the class as a
whole presumably based on the notion that an average assignment in respect to difficulty and amount will be helpful
to all members of the class.

        The study was concerned with three subordinate prob-
lems, as follows:

        (1)   Considering differences between boys and girls
and differences in ability in terms of achievement, how do
mastery test results of fifth-grade children who have

received individualized homework assignments compare with those of children who have received blanket-type assignments.

(2) With differentiation in respect to sex and levels of ability, what evidence can be established to compare the effects of individualized homework versus blanket-type homework insofar as pupil interest is concerned?

(3) How do the time requirements for pupils and teachers utilizing individualized homework assignments compare to the time requirements for pupils and teachers using blanket-type assignments?

Data were readily available in response to the first subordinate question simply by the stratification of experimental and control classes into subgroups by levels of ability. The stratification used in the experiment was accomplished by dividing the sample into approximately three equal parts in terms of the average scholastic achievement of the pupils. The dichotomy for boys and girls was achieved with equal facility and required no special alteration in the design used for the major problem.

Examination of the data in Table 1 shows that in two of the schools (B, C) almost no observed difference in mean gain is evident while marked differences in gain favoring the individualized treatment were found in the other two schools (A, D). When these data were subjected to the analysis of variance procedure, an F ratio of 9.07 for the treatment effect revealed that the difference favoring the experimental treatment was highly significant. However, it should be noted, as revealed in Table 2, that the difference in schools was highly significant, as was the interaction between the treatment and the individual schools. This significant interaction is interpreted by the investigator as evidence that the success of the individualized method is highly dependent upon the actions and reactions of individual teachers and pupils in the one to one relationship required in using this method. On the basis of this test, null hypothesis 1 was rejected.

All homework assignments throughout the experimental period were classified by teachers in terms of the primary purposes for which the assignments were given. These data reveal a fundamental difference between the blanket and individualized methods of homework assignment and are descriptive of the basic nature of the experimental method.

TABLE 1

Mean Gain And Variance For Experimental
(Individualized) And Control (Blanket-Type)
Classes In The Four Schools

School	Experimental Classes (Individualized)		Control Classes (Blanket-type)	
	Mean Gain	Variance	Mean Gain	Variance
A	33. 8	125. 6	26. 8	63. 6
B	46. 8	79. 4	47. 2	110. 0
C	36. 4	123. 7	37. 0	89. 8
D	42. 7	132. 6	31. 7	76. 8

TABLE 2

Significance Of Treatments For All Experimental
(Individualized) Versus All Control
Subjects (Blanket-Type)

Source of Variation	Sum of Squares	Mean Square	d. f.	F	Alpha . 05
School	7149. 3	2383. 1	3	23. 78*	2. 65
Treatment	908. 7	908. 7	1	9. 07*	3. 89
Interaction	1221. 0	407. 0	3	4. 06*	2. 65
Error	19435. 1	100. 2	194	. .	. .

*Significant at the . 01 level.

Homework, under the terms of the experiment, was ad-
judged as being either remedial, reinforcing (drill), or en-
riching in nature.   Results of the experiment reveal that
homework was exclusively of a reinforcement nature under
the blanket method of assignment, although this is not to
imply that remedial or enrichment assignments are not possible using this approach.   However, the results suggest
that homework designed primarily for remedial or enrichment purposes is apt to be quite limited or entirely omitted
under the blanket method of assignment.   Results with the
individualized method suggest different conclusions for that

treatment and those are the data with which we are primarily concerned in the discussion which follows.

The analysis of data classifying the individualized assignments is presented to disclose a fundamental difference between the two methods and also simply as descriptive of what occurred when the experimental (individualized) method was employed.   Results presented are not to be misconstrued as being a specific or desirable objective as revealed for the individual schools nor for the total experimental group.   A summary of these data reveals that 59.4 per cent of the individualized assignments were primarily for reinforcement purposes.   Similarly, 22.3 per cent of the individualized assignments were remedial and 18.3 per cent were designed primarily for enrichment purposes.

Findings

The statistical technique, analysis of variance, two-way classification was used to test null hypothesis 1.   This procedure enabled the researcher to determine if the difference between methods was significant and if the method was equally effective in the four participating schools.   The analysis showed:

(1)   A comparison of the mean gain in achievement on mastery tests between all subjects receiving individualized homework assignments and all subjects receiving blanket-type homework assignments showed significant differences favoring the group receiving individualized assignments as evidenced by a significant F ratio.   The gain was not observed in all participating schools and the difference among the schools was significant.

(2)   A comparison of the mean gain in achievement on mastery tests between boys receiving individualized homework assignments and boys receiving blanket-type homework assignments showed marked results favoring the group receiving individualized assignments in two schools and less pronounced results favoring the group receiving blanket-type assignments in two schools.   There was a significant difference among the schools but the difference between methods of assignment was not statistically significant.

(3)   A comparison of the mean gain in achievement on mastery tests between girls receiving individualized

TABLE 3

Mean Number Of Minutes Per Daily
Assignment By Levels Of Achievement
For Subjects Receiving Individualized Assignments
And Subjects Receiving Blanket-Type Assignments

Minutes	Level	Individualized Assignments		Blanket-type Assignments	
		No. of Pupils	%	No. of Pupils	%
Less 25	High	15	44. 1	16	45. 7
	Middle	9	29. 0	6	19. 4
	Low	5	14. 3	5	14. 3
Desired Range (25-34)	High	16	47. 1	17	48. 6
	Middle	17	54. 9	20	64. 5
	Low	24	68. 6	13	37. 1
35-more	High	4	11. 8	2	5. 7
	Middle	5	16. 1	5	16. 1
	Low	6	17. 1	17	48. 6

TABLE 4

Number Of Minutes Per Daily Assignment For Experimental
(Individualized Method) And Control
(Blanket-Type Method) Teacher

School	Individualized Method		Individualized Method	
	Mean No. of Minutes*	Range	Mean No. of Minutes	Range
A	14	**1 to 32	34	9 to 77
B	61	45 to 110	29	7 to 51
C	42	27 to 81	22	8 to 46
D	71	28 to 183	32	21 to 55
All Schools	47	1 to 183	29	9 to 77

*Does not include time when assignment was corrected or
determined in the classroom.
**One minute indicates assignments corrected and prepared
during class time with pupil assistance.

homework assignments and girls receiving blanket-type homework assignments showed gains favoring the girls receiving
individualized assignments in all four of the participating
schools.   The difference was statistically significant as was
the difference among the schools.

(4)   A comparison of the mean gain in achievement
on mastery tests between high achievers individualized homework assignments and high achievers receiving blanket-type
homework assignments showed gains favoring the group receiving individualized assignments in all four participating
schools.   The difference was statistically significant as was
the difference among the schools.

(5)   A comparison of the mean gain in achievement
on mastery tests between middle achievers receiving individualized homework assignments and middle achievers receiving blanket-type homework assignments revealed marked
gains favoring middle achievers receiving individualized assignments in two schools, marked gains favoring middle
achievers receiving blanket-type assignments in the third
school and little difference in the fourth school.   The difference between methods was not statistically significant for
middle achievers, however, there was a significant difference among the schools.

(6)   A comparison of the mean gain in achievement
on mastery tests between low achievers receiving individualized homework assignments and low achievers receiving
blanket-type homework assignments revealed differences favoring the group receiving individualized assignments in two
of the schools and little difference in the other two schools.
The difference between the methods was not statistically significant, however, the difference among schools was significant.

A Chi square test of independence was used to test
null hypothesis 2.   The analysis showed:

(7)   Interest in mathematics, as measured by pupil
choices at an interest table, was not significantly different
when the proportion of mathematical choices by all subjects
receiving individualized assignments was compared to the
proportion of mathematical choices by all subjects receiving
blanket-type homework assignments.

(8)   Interest in mathematics, as measured by pupil

choices at an interest table, was significantly greater for boys receiving individualized homework assignments than for boys receiving blanket-type homework assignments.   This interest, based on the proportion of mathematical choices, was greater at the conclusion of the experiment than at the mid-point of the experiment.

(9)   Interest in mathematics, as measured by pupil choices at an interest table, was significantly greater for girls receiving blanket-type homework assignments than for girls receiving individualized homework assignments.   This interest, based on the proportion of mathematical choices, was greater at the conclusion of the experiment than at the mid-point of the experiment.

(10)   Interest in mathematics, as measured by pupil choices at an interest table, was not significantly different when the proportion of mathematical choices by high achievers receiving individualized homework assignments, was compared to the proportion of mathematical choices by high achievers receiving blanket-type homework assignments.

(11)   Interest in mathematics, as measured by pupil choices at an interest table, was not found to be significantly different when the proportion of mathematical choices by middle achievers receiving individualized homework assignments was compared to the proportion of mathematical choices by middle achievers receiving blanket-type homework assignments.

(12)   Interest in mathematics, as measured by pupil choices at an interest table, was not found to be significantly different when the proportion of mathematical choices by low achievers receiving individualized homework assignments was compared to the proportion of mathematical choices by low achievers receiving blanket-type homework assignments.

Other Findings

(13)   Teacher control of the number of minutes of pupil time per assignment, showed little over-all difference when all subjects receiving individualized assignments were compared with all subjects receiving blanket-type assignments.

(14)   Teacher control of the number of minutes of

pupil time per assignment for boys showed results favorable
to the individualized method.   With the individualized approach, 60. 9 per cent of the boys were within the desired
time range as contrasted with 45. 6 per cent of the boys receiving blanket-type assignments.   Only 6. 5 per cent of the
boys were above the objective time range with the individualized approach as contrasted with 28. 3 per cent of the boys
receiving blanket-type assignments.

(15)   Teacher control of the number of minutes of pupil time per assignment for girls showed an almost identical
pattern of girls above, below, or within the desired time
range under the two methods of homework assignment.

(16)   There were no marked differences in teacher
control of pupil time per assignment for high achieving pupils between the two methods of homework assignment.   Almost one of every two high achievers required less time
than anticipated by the teachers using the blanket-type method (45. 7 per cent) but a similar result occurred with teachers using the individualized method (44. 1 per cent).

(17)   There were no marked differences in teacher
control of pupil time per assignment for middle achievers
between the two methods of homework assignment.

(18)   Marked differences favored the individualized
method of assignment for low achievers.   Almost half (48. 6
per cent) of the low achieving pupils receiving blanket-type
assignments required more time than the desired time range
as contrasted with the 17. 1 per cent of low achievers receiving individualized assignments.   Additionally, 68. 6 per
cent of the low achievers were in the desired time range
under the individualized method as contrasted with only 37. 1
per cent for the blanket-type method.

(19)   The mean number of minutes of out of school
teacher time required to prepare and correct the individualized assignments was 47 minutes per assignment as contrasted with 29 minutes per daily assignment for teachers
using the blanket-type method of assignment.

(20)   Analysis of the blanket-type assignments of the
experiment revealed them to be entirely of a reinforcement
nature.

(21)   Analysis of the individualized assignments of the

experiment revealed that 59. 4 per cent were of a reinforcement nature, 22. 3 per cent were for enrichment purposes.

Notes

1.   Department of Public Instruction.   The Elementary
      Course of Study.   Bulletin 233-B, An Interim Report.
      Harrisburg: Commonwealth of Pennsylvania Depart-
      ment of Public Instruction, 1949; p. 56.

2.   Ruth Strang.   Guided Study and Homework: What Re-
      search Says to the Teacher.   No. 8.   Washington,
      D. C. :   Department of Classroom Teachers, American
      Educational Research Association, National Education
      Association, 1955; also Ruth Strang, "Helping Your
      Child With His Homework, " The PTA Magazine 56:24-
      27, November 1961.

3.   Avram Goldstein.   "Does Homework Help?   A Review
      of Research. " Elementary School Journal 60:212-
      224, 1960.

INDIVIDUALIZED VERBAL ARITHMETIC PROBLEM
ASSIGNMENTS

Cecil T. Nabors

       The purpose of this study involving 316 fifth-grade
pupils in five elementary schools was to determine the effect
of individualized problem-solving assignments on the achievement of fifth-grade pupils in mathematical concepts and problem solving.   One-half of the members of each class were
selected at random to serve as the experimental group.   The
other members of the 12 classes served as the control group.
The arithmetic concepts and problem-solving sections of the
Iowa Tests of Basic Skills were administered as a pretest.

Materials and Procedures

       Materials used by the experimental group of children
were verbal problems selected from the elementary

mathematics textbooks used in the public schools of the state of Texas. In order to identify the level of difficulty, verbal problems taken from, for example, the first part of second-grade books were labeled 21, and the problems taken from the last part of the second-grade books were labeled 22.

In a similar manner problem-solving assignments were prepared and labeled for two levels at the third, fourth, fifth, and sixth grades, and one level at the seventh grade. The investigator assumed that the problems included in a textbook had the level of difficulty represented by the grade level of the textbook. The wording of the verbal problems was not altered except for changes of proper names so that the reading level was not changed. Fifteen sets of five problems each were prepared for both levels of third, fourth, fifth, and sixth grades, and ten sets were prepared for the other levels.

Some details of the materials used in the study were formulated and tested in a pilot study conducted during the fall of 1965.

The readability level of each verbal problem level was checked by use of readability formulae. The readability level matched the level designated as the problem level in 90 per cent of the problems. The reading level of verbal problems should be lower than material used for reading instruction. Pupils should not be limited in their efforts to solve verbal problems because they cannot read the conditions stated in the problem. Most of the materials used in this study were more difficult to read than the labeled grade level indicated.

It was determined that the low level problems were too easy for pupils having the corresponding grade equivalent score on the Iowa Tests of Basic Skills, and the higher level problems were too hard for pupils with the corresponding grade equivalent score on the standardized test. If a pupil had a grade equivalent score in problem solving of third grade or below, raising this score by one grade provided the appropriate level for problem-solving assignments. If the pupil had a grade equivalent score of seventh grade or above, a sixth-grade assignment was appropriate.

The proposed study was discussed with principals in each school and individual conferences were held with the fifth-grade teachers agreeing to participate in the study.

After each child had completed two of the assignments, an analysis of the results was made and a decision
made as to the necessity of changing his assignment to
easier or more difficult problems.   An effort was made to
provide problem-solving assignments that were not too difficult nor too easy.   Each pupil was allowed 15 minutes to
work independently on his problem-solving assignment.   After finishing the assignment, the pupil checked his work
with the appropriate answer sheet.   The pupil made a record of his correct answers on his personal record sheet
and used the answer sheet in his study of the problems that
he failed to solve.   The pupils of the experimental group
worked the individualized problem-solving assignments three
days a week for a period of ten weeks.   Fifteen sets of
problems were selected for most of the levels.   Most of the
pupils worked at one level for five weeks and then worked
at a higher level for the last five weeks of the experimental
period.   The teacher in some cases varied this procedure
for certain individuals.

During the ten weeks the experimental group worked
problems at their ability level, the others worked with regular fifth-grade textbook materials.   Each teacher kept a log
of the time each group spent on verbal problems and planned
pupil assignments so that the experimental group did not have
the experimental problem-solving assignments in addition to
assignments given the whole class.

The teachers involved in the study differed in their
control of the time each group spent on verbal problems.
Some were able to balance the time, but other teachers did
not provide as much time on verbal problems for the control
group as the experimental group used.   Using the log of
time spent on verbal problems kept by each teacher, the
total time spent on verbal problems by the two groups in
each class was determined.   The control group as a whole
averaged approximately one-half as much time on verbal
problems as did the experimental group.

After ten weeks all the pupils were given another
form of the arithmetic section of the Iowa Tests of Basic
Skills.   The raw scores of the final test minus the raw
scores of the pretest provided score gains for each pupil in
the two areas of mathematics.   For analysis the members
of the experimental group and the control group were separated into subgroups based on sex, reading ability, and intelligence.   The grade equivalent score in reading on the Iowa

Tests of Basic Skills and the intelligence quotient determined by the Otis Mental Ability Test: Beta Test were used in the classification of pupils into three subgroups for reading and for intelligence.

Results

With the score gains of the arithmetic concepts test as the criterion measure, tests of significance failed to reject any of the null hypotheses of no difference between the experimental group or subgroups and the control or subgroups.

With the score gains of the problem-solving test as the criterion measure, the null hypothesis of no difference in score gains of the pupils with average intelligence using individualized verbal problem assignments and score gains of pupils with average intelligence using regular fifth-grade materials was rejected with a 5 per cent level of confidence. Tests of significance failed to reject any other null hypothesis involving problem-solving score gains as a criterion.

Conclusions

Score gains in problem-solving by pupils of average intelligence using individualized problem-solving assignments were significantly greater at the 5 per cent level of confidence than the score gains of pupils of average intelligence using regular fifth-grade mathematics textbook materials.

MASS METHOD VERSUS INDIVIDUAL METHOD
IN TEACHING MULTIPLICATION
TO FOURTH-GRADE PUPILS

Maria Paul Morrison

In a minor research problem in four fourth-year grades an attempt was made to determine whether multiplication is more effectively taught by the mass method of instruction or by the individual method.

Four classes were used in the experiment, two being

taught by the mass method and two by the individual method. There were 62 children in the former group and 75 children in the latter group.

The teaching of the mass method included the use of flash cards of the basic multiplication facts, games, the rainbow device for crediting pupils with table mastery, and written practice with corrections by the teacher.

The teaching of the individual method included the use of a drill book, the use of individual study cards of errors made in basic facts, and pupil checking of work with final checking of practice by the teacher.

The four classes maintained their heterogeneous grouping with relatively little difference in median intelligence quotients and chronological ages.   The median intelligence quotient of the mass method group was 105 and of the individual group 103.   The chronological age of the former group was 9 years 9 months while that of the latter was 9 years 5 months.

In the experiment the Wilson Process Test in Multiplication, 5P, was used as a pretest in October, as a final test in March and as a retest in September.   Forty-seven periods of 45 minutes were devoted to the teaching of multiplication.   Reliance was placed on motivated drill.

Comparison Of Medians Of Pretest, Final Test, And Retests By Two Methods, Mass And Individual

	Mass	Individual
November Pretest	18.00	8.27
March Final	84.83	80.12
September Retest	64.00	68.00
Immediate Gain (at end of experiment)	66.83	71.85
Permanent Gain over Pre-test (in Sept. before review period)	46.00	59.59
Summer Losses	20.83	12.12

The median scores of the three tests are listed in

the accompanying table. From this table it is noted that
using the mean as a basis for comparison the group taught
by the individual method using the Wilson Drill Book made
a slightly greater gain than the group taught the nine tables
in the traditional manner. Comparing the September mean
on the retest with the initial mean of the pretest it is also
evident that the permanent gain of the individual group was
13.59 points higher than the group taught by the class method. When the September mean is compared with the March
mean it is seen that the summer losses of the mass method
group were 20.83 points and of the individually taught group
were 12.12 points.

In considering the slight advantage of the pupils taught
by the individual method over the other group the difference
is in all probability due to at least three outstanding factors
in the teaching methods, which are motivation, checking,
and written work.

In the group taught by the individual method the motivation was pupil induced rather than teacher-induced. When
the work of the pupils is pupil driven rather than teacher-
driven then there is more likely to be increased accomplishment, energy and power.

The second factor of significance was in the checking.
In the mass group the teacher checked the work of the pupils, calling attention to errors and corrections. In the
other group the pupils did the first checking and attempted
to prove all examples by reversing multiplicant and multiplier. The latter group worked until every result was 100
per cent correct. The teacher did the final checking of the
drill book exercises.

The third factor which enabled the individual group
to acquire more permanent learnings was the use of the
drill book material which eliminated errors in copying,
saved time, thus permitting the pupils to do more practice
work in multiplication in the forty-five minute period.

As a result of this attempt to evaluate the two methods on the basis of pupils' achievement the fourth-grade
teachers felt that insufficient time was devoted to the teaching of multiplication for mastery, and that process drill
books are a desirable means of individual instruction.

X. INDIVIDUAL DIFFERENCES

This chapter is designed to present selections that
are concerned with the individual differences that exist among
children in mathematics classes.   In a sense, much of the
material preceding this chapter has been based on the premise that the existence of individual differences is recognized
by the reader.   Nevertheless, it seemed to me to be appropriate to emphasize this point once more.   Some of the
items that I have selected for inclusion in this chapter do
not concern themselves with the individualized procedures
that form that subject matter of this book and it is important
that they not be read in order to obtain information about
that aspect.   It is equally important, however, to recognize
that each of the authors of these articles emphasizing the
existence of individual differences also recognizes that this
emphasis must result in a modification of lock-step instruction.

R. Stewart Jones and Robert E. Pingry

The first item in this chapter is a reprint of the first
half of a longer article which first appeared in the 25th Yearbook of the National Council of Teachers of Mathematics,
published in 1960.   I wholeheartedly recommend this article,
as excerpted here and as a whole in its original source, as
one of the best concise statements ever to appear in the
literature on the individual differences that exist among
learners.

The article is written by two professors at the University of Illinois in Urbana who are interested both in
mathematics and the field of education.   It is deserving of
very careful reading by each individual who is interested,
either as a researcher or as a practitioner in a classroom.
There is much in this well-organized item that will add to
the knowledge and understanding of every member of these

two groups.

Florence Flournoy

Professor Flournoy of the University of Texas is the author of the original article from which the second part of this chapter is excerpted.   In this item a discussion and description of ways in which individual differences can be accommodated are given.   It is interesting that individualization is mentioned as one of the possible solutions but apparently was not included by any of the teachers in the project described.   I have mentioned in several places in these chapter introductions that individualization is not a panacea, rather that it is or should be employed only when teachers are ready and receptive to its use.   There can be little doubt that the testing of various other approaches toward meeting individual differences may well be not only a desirable but often a very necessary step toward the introduction of individualized instruction at a later time.

My reason for the inclusion of this article is that it clearly delineates a situation in which there is recognition of the existence of individual differences among learners as well as of the need for adapting instruction so that differences will be taken into account; and there is an attempt to meet the problem squarely on the part of classroom teachers who are flexible enough to try various procedures in their own classrooms over a period of time long enough to allow evaluation concerning the comparable merits of each procedure.

Oscar T. Jarvis

The last item is taken from a journal article written by Professor Jarvis, chairman of the Department of Curriculum and Instruction of the University of Texas at El Paso. The portion of the article which I selected for inclusion in this volume gives the results of an analysis of individual differences in a number of arithmetic classes.   The analysis makes very clear the relative degree of independence that exists between intelligence and school achievement in a particular subject.   Certainly we cannot possibly predict one from our knowledge of the other with any fair degree of accuracy in the case of any one individual pupil.   This fact shows how futile it is to attempt to do away with the necessity of accommodating to individual differences by putting children into so called "homogeneous" classes on the basis of scores on an intelligence test.

Professor Jarvis is very explicit in pointing out that while it is extremely important that there be a recognition and acceptance of the existence of individual differences by teachers, it is even more essential that something be done, not to eliminate the differences, for that is impossible, but to provide for meeting the instructional needs that occur because these differences exist.

INDIVIDUAL DIFFERENCES

R. Stewart Jones and Robert E. Pingry

That children of the same age differ widely in the mental, physical and personal traits which affect schooling is now common knowledge[2, 26]. Educators and psychologists have portrayed so well the diversity among children and have so emphasized the concomitant problems of teaching that many teachers have become convinced that adequate handling of so many different children in one class is impossible. Furthermore, the widespread attention to this topic has often carried with it the unfortunate implication that differences among pupils can and should somehow be ironed out. Nothing could be further from the truth. Differences among children will increase as a result of development and the school should have a hand in this development by making children more different, rather than less. The problem, then, is not that individual differences exist, but what can be done in teaching to take account of these differences.

With nearly 100 per cent of elementary age children in school, with rapidly increased birth rates since the last war, and with the holding power of the high school increasing each year[1], teachers should anticipate not only more pupils but also greater diversity among them than was the case even a decade ago. Children are coming to school younger than ever before, they are staying in school longer than ever before, and they are remaining in classes with their chronological age mates more than in the past[14].

Furthermore, there are far too few special classes and remedial staffs to handle even the extreme deviates[1]. All of the foregoing facts do not point to a new problem; they focus sharply upon an old problem which few schools have tackled with sufficient imagination and vigor[11].

General Psychological Considerations

The number of human traits is unknown, but undoubtedly there are thousands. Cattell[5] has attempted to differentiate between the innumerable surface traits which are apparent in behavior and personality, and source traits, which give rise to manifest behavior. Similarly, Thurstone[23] has developed methods for identifying the separate primary mental abilities. Both psychologists have attempted to group manageable and meaningful traits to describe individuals and their differences. Nevertheless, a composite description of a child, even using the factors of personality and ability, presents a complex picture. Every child is unique.

But children of a given culture also have much in common. Cattell[5] attributes many of our similar traits to what he calls "environmental mold unities" and Gesell[10] has emphasized the many similarities in the sequential and orderly progression of development. If each child is seen as completely unique, it follows that instruction must be completely individualized; if all children are viewed as alike, teaching should be entirely a group process. Neither position is tenable, and all good teaching accepts a compromise somewhere between these extremes. Many traits can be measured. Children can be identified as possessing more or less of a characteristic that affects schooling, and the general characteristics of a class can be described. Clearly such measurement and the description that follows are only approximations but they are extremely useful. Measurement has provided the knowledge that we have about differences among children and has given clues as to their causes.

The following propositions about individual differences should be a starting point for devising efficient educational procedures to handle them; these generalizations will later be related to the problems of teaching and of school organization:

(1) Children in any classroom, even in those where ability grouping is used, will differ widely in both general

and special abilities and aptitudes[3].

 (2)   Differences in both ability and achievement increase with age, the span or range of general ability increasing almost twofold from the first to the eighth grade[7].

 (3)   The various special abilities (for example, numerical fluency) correlate differentially with each other and with any measure of general ability that is presently in use[16].

 (4)   Any school subject requires a variety of abilities[16].   Competency in arithmetic, for example, calls for numerical fluency, numerical comprehension, conceptual ability, visual and auditory memory, etc.

 (5)   Differences between the sexes in abilities and skills are slight, with some apparent advantage for the earlier maturing girl, especially in language skills, and some slight advantage for boys in mathematics[2].

 (6)   Differences within an individual, that is, the variability of traits within a single child, are more than half as great as differences among children[7].

 (7)   Although various abilities overlap, that is, intercorrelate, they are sufficiently independent to merit separate measurement[16].

 (8)   Differences not accounted for by mental age measures are as important in determining school success as intellectual ones.   One-fourth of retarded readers, for example, are above average in measured intelligence[8].

 (9)   The effects of school that fails to consider differences among children seem to be cumulative.   Children fall further and further away from the attainments of which they are capable at all levels of ability, but particularly at the higher ability levels[1].

 (10)   Differences are not static.   Children vary when they enter school and they develop at different rates.   Consequently, diagnosis must be a continuous process.

Differences In Arithmetical Skills

 Theoretically, differences in arithmetic achievement might be expected to vary as greatly as intelligence or as aptitude for arithmetic varies.   Actually the variation is probably somewhat greater.   The correlation between intelligence (or any of its factors or combinations thereof) and arithmetic achievement rarely exceeds .60 and in most cases proves to be lower.   Obviously, unmeasured factors such as experience, emotional stability, modes of thinking, and

attitudes contribute to success in arithmetic. The extent of
the variability is portrayed in Table 1. As may clearly be
seen there, many children are far beyond the comptence
that might erroneously be thought of as the standard for

TABLE 1

Grade Level Scores in Arithmetic Earned by 519 Eighth-
Grade Pupils[1]

Grade level (by pupils' scores)	13	12	11	10	9	8	7	6	5	4
No. of children at that level	11	14	60	142	134	81	47	21	8	1

---

eighth-grade work. Witness the 25 children who were already as advanced as high school seniors or college fresh-
men. Similar findings by Frandsen[9] in five sixth-grade
classes in Logan, Utah, showed 47 of the 195 children
tested to be at a grade-equivalent at least one year ahead
of norms. Such evidence dramatically portrays the problem
of trying to teach the same arithmetic lesson to every pupil
in a given class. However, on a scale of total achievement
one gets neither sufficient diagnostic information to take
specific action nor a clear picture of what the differences
mean in operational terms. More relevant information on
the meaning of such differences has been cited by Brownell[4],
who tested 487 children in 20 fifth-grade classes in the component skills required for performing the operation of division. All these children were about to embark on this
relatively complex arithmetical task. The results not only
gave evidence of the expected wide range of achievement
(half were not ready to begin this study) but also and more
important, identified the specific areas of weakness of each
child.

        Little can be gained at this point by further elabora-
tion of the range of achievement in arithmetic. Any school
that keeps records and uses them effectively has ample evidence of its own. Suffice to say here that such differences
are not limited to the lower grades nor are they limited
alone to computational skills. At every age level there is
great variability in all areas of arithmetical and mathematical skill and understanding.

More important for this discussion is the fact that in the upper grades, in high school and college, and in adult life there are large numbers of people who do not possess even the minimum essentials of arithmetical skill, basic mathematical concepts, and related applications to everyday experience. The writer recently asked 124 graduate students what the diameter of the moon is. The answers ranged from one mile to 10 billion miles. Similar evidence of inability to use mathematical concepts is given by Horn[12], who cites a half-dozen studies to support this proposition. An anecdote in this regard is furnished by Richter[20], who describes the case of a teacher who asked a fourth-grade class to find the product of .08 and 1/5. The teacher was berated by the father of one of the children; he had given up trying to solve his daughter's problem when told by a successful business acquaintance and a Spanish professor that the problem was impossible.

The school's goal, then, is not to reduce differences among children, but to take account of them in teaching and to attack the problem of widespread ignorance presently existing in this area. To this problem we now turn our attention.

Factors of Individual Differences (and Educational Casualties)

The factors that produce variability among pupils and that in some cases result in educational casualties include a variety of both innate characteristics and environmentally-produced traits. For purposes of the present discussion these factors are divided into three groups. Group I includes those basic factors over which the school has little control; Group II embraces those conditions arising outside the school that produce differences, but over which the school may have an ameliorating influence; and Group III contains those causes of difference that arise in the school itself.

Group I Factors. The school has practically no control over, but must make adequate teaching provisions for these factors, which include innate potential, accidents of birth and environment, and irreversible organic changes. Mental potential, one of the most important as a determiner of learning, is portrayed in Table 2.

As may be seen in Table 2, four-year-olds would be

TABLE 2

Theoretical Distribution of Mental Ages for Groups of 100
Children with Given Chronological Ages
(Based on a theoretical standard deviation of IQ of 16. 6)

Mental Age	Chronological Age										
	4	5	6	7	8	9	10	11	12	13	14
20											1
19										1	2
18										1	4
17									1	4	8
16								1	3	7	12
15								2	7	13	15
14							2	6	12	16	16
13						1	5	12	17	17	15
12					1	4	12	18	19	16	12
11					3	11	19	21	17	13	8
10				2	11	21	23	18	12	7	4
9			1	9	22	25	19	12	7	4	2
8			7	23	28	21	12	6	3	1	1
7		5	24	31	22	11	5	2	1	1	
6	2	24	36	23	11	4	2	1			
5	23	41	24	9	3	1					
4	50	24	7	2	1						
3	23	5	1								
2	2										
Totals	100	99	100	99	102	99	99	99	99	101	100

expected to vary from two to six years in mental age; at
eight they would vary from four to 12 years, and at 14,
from eight to 20 in mental age, a 12-year span.

About 15 per cent of all children in public schools
have IQs between 75 and 90, and are sometimes referred to
as the "dull-normal" group. These children should not be

thought of as mentally handicapped, nor can one find much
evidence to argue for separate classes for them.   Because
of the already noted intra-individual variability, some of
these children will be at or above average in some of the
components of arithmetical skill.   As a group, however,
they do constitute a serious problem for the school which
uses an inflexible grade placement of topics in arithmetic
or in any other subject.   Another 4 to 5 per cent of the
school-age group may be classified as mentally handicapped;
that is, with IQs between 50 and 80.   In one of our most
progressive states only 16 per cent of this group is cared
for in special classes.   Undoubtedly, throughout the country
as a whole, most of this group is in the regular classroom.
Children whose measured IQs are below 50 are generally
referred to as trainable rather than educable.   Actually,
however, many of these children are also to be found in the
regular classroom.

At the other extreme, 1 to 5 per cent of the children
(depending on how defined) may be thought of as gifted or
talented.   In many cases the educational attainments of these
children are far below the capacity of which they are capable[1].

Intra-individual variability, as already mentioned,
further complicates the problem of caring for individual differences.   In one study 25 children, all with IQs of 106,
varied in mental age on different tests from one to almost
eight years in such areas as memory, space perception, and
verbal ability[24].   In view of this type of diversity, Traxler[25]
has emphasized the need for testing that yields a profile of
abilities.   He suggests such tests as the Primary Mental
Abilities Test, the Differential Aptitudes Test, and the Yale
Educational Aptitudes Test as measures that will provide
useful information for educational planning.

Beside mental abilities there are other conditions,
such as physical handicaps and defective vision and hearing,
that affect schooling.   Only a small percentage of this latter
group receives adequate special services, and consequently
these children become a problem for the regular classroom
teacher.

For all the Group I factors the school can do little to
effect a change.   It is futile to believe that even the most
excellent teaching can create readiness for learning among
those children whose mental potential is far below that of

their classmates.   The only recourse is to make appropriate adjustments in both curriculum and teaching methods.   Suggestions for ways to handle these innate or uncontrollable differences will be discussed shortly.

Group II Factors.   These factors arise outside the school, but are products of learning and maladaptation and are amenable to treatment within the school.   Included are unfavorable attitudes, emotional disabilities, fears and anxieties, motivational differences and erroneous modes of thinking.   These factors may be general personality characteristics, such as anxiety, which interfere with all school work, or they may be specific attitudes about arithmetic which arise in the home or community.   One researcher found that children who are superior readers dislike mathematics more often than their peers[21].   It is well known that girls, who are more proficient than boys in language skills, have less interest and ability in arithmetic.   Girls rank arithmetic or mathematics in about seventh place when asked what school subjects are most valuable, while boys characteristically place mathematics in first place.

The extensive effect of Group II factors may be seen in the fact that one-fourth of the children who make slow progress in school are of normal or superior intelligence[8]. Even the community in which a child lives may bear a significant relationship to his success in school.   Martens[17] found that comparable groups of urban and rural children were significantly different in school achievement with differences in both reading and arithmetic favoring the former group.   Incidentally, in this same study there was a greater difference in measured school achievement in arithmetic than there was in reading.

That the school is in a good position to ameliorate the damaging effects of some of the out-of-school deterrents to achievement has been demonstrated in a "total push" program described by Coleman[6] in which children were given intensive help for a period of only six weeks.   The steps used in the program encompassed not only the creation of a favorable atmosphere for learning but also an attempt to integrate the home environment with the school program. Gains were made in all school subjects with greatest gains in the area of arithmetic.   Both Coleman's study and the previously cited study by Martens suggest that the area of arithmetic is especially sensitive to Group II factors.

The general effects of factors such as attitude and out-of-school experience have been seen to have sufficient impact to alter measured intelligence.   As McCandless[15] has pointed out, some children develop generalized attitudes unfavorable toward learning almost anything the school has to offer, and other children acquire early habits of thinking in concrete rather than abstract terms.   Such children are badly handicapped.   Even more severely handicapped are those children who have been afforded "the richness of opportunity to learn self-defeating behaviors"[15] (p. 684).

Group II factors operate to depress and inhibit the realization of full potential at all levels of ability and at all ages.

Group III Factors.   These factors arise in and because of the school.   They include poor teaching, poor understanding of arithmetic and mathematical concepts and adverse attitudes about arithmetic on the part of teachers, inflexible and unwise grade placement of arithmetical topics, failure to take account of pupils' readiness for learning, and failure to build arithmetical readiness in the early grades.

Several writers[11, 18] have shown that many teachers do not have either sufficient understanding or sufficient skill in arithmetic.   Orleans and Wandt[18] tested more than 100 teachers on their understanding of such simple arithmetical processes as multiplication.   One-half of the teachers' answers indicated incomplete or almost total lack of understanding of such processes.

For example, almost one-half of the teachers missed the following: "Look at the example at the right.   Why is the third partial product moved over two places and written under the 2 of the multiplier?"

$$
\begin{array}{r}
157 \\
246 \\
\hline
942 \\
628 \\
314 \\
\hline
38622
\end{array}
$$

As noted by Brownell[4], when a given topic is arbitrarily placed at a given grade level, many children will have neither the necessary mental equipment nor the prerequisite skills to allow satisfactory progress without special help.   In most cases this special help entails reteaching of earlier skills.   Teachers are well aware of the problem of the schools' need to consider individual differences, but many lack knowledge of and skill in the necessary techniques and methods for dealing with these differences[15].   Lee maintains that our failure in this regard stems from a variety of factors including lack of adequate knowledge of children, unsufficient utilization of materials that are available, and failure to use what is known about teaching small groups. Apropos here is Swenson's[22] analogy in which she compares progress in arithmetic by two pupils of very different abilities with distances covered by two cars of different makes and speed. The effect of a single road and a single speed is a reduction in the efficiency of both cars. Similarly, she asserts, two very different children who are forced to travel through the same exercises at the same rate of speed are both unable to realize the achievements of which they are capable.

Finally, the school must assume a major share of the responsibility for failure to capitalize upon many opportunities for building children's readiness for arithmetic. When formal arithmetic instruction begins, especially drill work, many children have not yet acquired the necessary equipment, attitudes and experiential background. Consequently their initial experiences with number work are unpleasant and they may build avoidance reactions to arithmetic. Perhaps even more unfortunate, pupils may at this point form an adverse attitude toward problem-solving in general. The effectiveness of a program to build readiness has been demonstrated in a study by Koenker[13], who compared kindergarten children who had received a readiness program with other kindergarten children who had not. The readiness program consisted of numerous activities such as measuring the room and objects in it, counting, and simple number games. The group which received the early experience with arithmetical concepts were significantly higher on an arithmetic readiness program consisted of numerous activities such as measuring the room and objects in it, counting, and simple number games. The group which received the early experience with arithmetical concepts were significantly higher on an arithmetic readiness test than the control kindergarten section.

Special efforts to diagnose pupil difficulty and to help pupils learn, as has been so richly demonstrated for reading, may bring about dramatic improvement in arithmetic by alleviating the mistakes previously made in children's early school experiences. Of all the factors that produce differences among children, those which result from poor teaching and programming should be the first to come under attack by arithmetic teachers everywhere.

Methods of Handling Individual Differences

      That individual differences exist is now apparent.
Experienced teachers need not be told of the great range
and variety of differences.   Teachers are constantly being
reminded of these differences as they teach, and they are
constantly seeking methods for dealing with them.   Few
teachers, however, are ever satisfied that they are doing a
good job in this respect.

      At professional meetings and in teachers' discussion
groups, the subject of how to care for individual differences
is a frequent one, and one of obvious concern.   Teachers
want and need assistance in finding procedures that will enable them to help a greater percentage of their pupils obtain
a satisfactory educational experience.

      Teachers of arithmetic, administrators, and educators
are not agreed upon the policies and procedures for caring
for individual differences.   In fact, on some of the policies
and procedures very active arguments take place in faculty,
PTA, and school board meetings.   Some of the major issues
on which the arguments are presented are the following:

      (1)   Some persons are active in supporting a horizon-
tal enrichment of the able pupil's experience in mathematics.
These persons are opposed to advancing the pupil to more
abstract mathematics or topics that are usually taught in
succeeding grades in school.   They claim it is possible to
provide adequate and desirable learning experiences for the
able pupils without advancing vertically.   Other persons are
strong in supporting the vertical advancement of the able
pupils to more advanced mathematics.   They believe the
failure to advance causes boredom and poor work habits on
the part of the able pupils[19].

      (2)   Some persons are opposed to the homogeneous
grouping of pupils within a grade level.   They claim that
homogeneous grouping really is impossible, that it is non-
democratic, and that it is hard on a pupil's mental health.
Other persons are in favor of grouping pupils on the basis
of achievement and ability; they argue that mathematics is
a sequential subject that requires understanding of basic
ideas at one level before learning can proceed to a next
level.   They deny that homogeneous grouping is non-demo-
cratic or a cause for poor mental health.

(3)   Another basic issue concerning the problem of individual differences has to do with promotion policy. Some persons urge that promotion in school be based almost solely on achievement results. They blame "social promotion" as a major cause for the wide dispersion of talent in some of the upper grades of the elementary schools. Others are in favor of a promotion policy that includes achievement as one of the factors, but also includes social age-level adjustment as one of the major factors.

(4)   Another issue concerns strict adherence to the grade placement of topics. Some persons are in favor of setting standard grade level objectives in mathematics and then requiring that all pupils meet these standards. The teacher should not go beyond these standards, they state, because of the administrative difficulties that result. Other teachers and administrators favor rather flexible grade level standards, set according to the ability of the class. In some cases this means that the class will proceed to a study of more advanced topics, and in other cases it will mean that the class will not complete the usual work of some standard curricula in mathematics.

(5)   There is a polarity of views toward the type of study that should be provided the able pupil in mathematics. Some persons favor supplementary work for these pupils whereby they would gain experience in the social and economic applications of mathematics. They maintain that school experience contains too little reference to the relationships among the major disciplines and the ramifications in general education. Others believe that social-economic applications are both inappropriate and boring for the bright pupil, who needs work in abstract mathematics.

Very little definite research exists on the above arguments, and teachers will not find it possible to choose a position solely on the basis of research findings. Some of these issues probably can never be settled on a purely empirical basis, either, for basically the arguments are philosophical in nature. Experiments and surveys would be of considerable help in providing facts, but the final decision rests upon assumptions made in one's educational philosophy.

Some school faculties and administrations have well-defined policies, and the individual teacher can work as a member of the faculty to modify or support these policies. However, the principal approach of the teacher will be to adjust to existing school policy and work out the best program of instruction within this policy.

Whatever the over-all school policies, it is still the individual classroom teacher who must adapt instruction to the differences he finds among his pupils. The teacher can be encouraged by the knowledge that there is a growing variety of instructional materials to assist him in his task. Finally, it should be heartening to note the growing awareness of this problem among both professional and lay groups.

References

1. Allerton House Conference On Education. (Edited by R. G. Bone and R. Stewart Jones.) The Nature of the School Population in the State of Illinois. Bulletin No. 24. Springfield, Ill.: Illinois Curriculum Program, Office of Superintendent of Public Instruction, June 1955. [Table 1 adapted from this source.]

2. Anastasi, Anne, and John P. Foley. Differential Psychology. New York: Macmillan, 1949.

3. Blair, Glenn M., R. Stewart Jones and Ray H. Simpson. Educational Psychology. New York: Macmillan, 1954.

4. Brownell, William Arthur. "Arithmetical Readiness as a Practical Classroom Concept." Elementary School Journal 52:15-22, September 1951.

5. Cattell, Raymond Bernard. Description and Measurement of Personality. New York: World Book, 1946.

6. Coleman, James Covington. "Results of a 'Total-Push' Approach to Remedial Education." Elementary School Journal 53:454-58, April 1953.

7. Cook, Walter Wellman. "Individual Differences and Curriculum Practice." Journal of Educational Psychology 39:141-48, March 1948.

8. Durrell, Donald D. "Learning Difficulties Among Children of Normal Intelligence." Elementary School Journal 55:201-208, December 1954.

9.   Frandsen, Arden N. How Children Learn. New York:
     McGraw-Hill, 1957.

10. Gesell, Arnold, and Frances L. Ilg. Child Develop-
     ment. New York: Harper and Brothers, 1949.

11. Glennon, Vincent J. "A Study in Needed Redirection in
     Preparation of Teachers of Arithmetic. " Mathematics
     Teacher 42:389-96, December 1949.

12. Horn, Ernest. "Arithmetic in the Elementary School
     Curriculum. " The Teaching of Arithmetic. 50th
     Yearbook, Part II, N. S. S. E. Chicago: University
     of Chicago Press, 1951. Chapter 2, p. 6-21.

13. Koenker, Robert H. "Arithmetic Readiness at the Kin-
     dergarten Level. " Journal of Educational Research
     42:218-23, November 1948.

14. Lennon, Roger Thomas, and B. C. Mitchell. "Trends
     in Age-Grade Relationship: A 35-Year Review. "
     School and Society 82:123-25, October 15, 1955.

15. McCandless, Boyd. "Environment and Intelligence. "
     American Journal of Mental Deficiency 56:674-91,
     April 1952.

16. Manual for Functional Evaluation in Mathematics. Test
     Battery. (Edited by Ben A. Sueltz, and W. A.
     Brownell.) Philadelphia: Educational Test Bureau,
     1952.

17. Martens, Clarence C. "Educational Achievements of
     Eighth-Grade Pupils in One-Room Rural and Graded
     Town Schools. " Elementary School Journal 54:523-
     25, May 1954.

18. Orleans, Jacob S. , and Edwin Wandt. "The Understand-
     ing of Arithmetic Possessed by Teachers. " Elemen-
     tary School Journal 53:501-507, May 1953.

19. Pressey, Sidney L. Educational Acceleration: Apprais-
     als and Basic Problems. Columbus: Ohio State Uni-
     versity Studies, Bureau of Educational Research
     Monographs 31, 1949.

20. Richter, Charles O. "Readiness in Mathematics. "

Mathematics Teacher 37:68-74, February 1944.

21. Sheldon, William D. , and Warren C. Cutts. "Relation
    of Parents, Home and Certain Developmental Charac-
    teristics to Children's Reading Ability, II. " Elemen-
    tary School Journal 53:517-21, May 1953.

22. Swenson, Esther J. "Rate of Progress in Learning
    Arithmetic. " Mathematics Teacher 48:70-76, Febru-
    ary 1955.

23. Thurstone, Louis Leon, and Thelma G. Thurstone.
    Factorial Studies of Intelligence. Psychometric
    Monograph No. 2. Chicago: University of Chicago
    Press, 1941.

24. Tiegs, E. W. Tests and Measurements in the Improve-
    ment of Learning. Boston: Houghton Mifflin, 1939.

25. Traxler, Arthur E. "The Use of Tests in Differentiated
    Instruction. " Education 74:272-78, January 1954.

26. Tyler, Leona E. The Psychology of Human Differences.
    New York: Appleton-Century-Crofts, 1947.

MEETING INDIVIDUAL DIFFERENCES IN ARITHMETIC

Frances Flournoy

That children vary widely in their interest and
achievement in arithmetic is an accepted fact. A variety
of factors influence pupil interest and achievement in arithmetic and these factors must be taken into consideration in
planning ways of helping each individual to progress in arithmetic in accordance with his ability. A type of class organization must be selected which will facilitate the carrying
out of variations appropriate to meet the varying needs of
individuals.

Ways of Varying Instruction

Believing that making certain instructional variations

is essential when making plans for meeting individual differences, it has been proposed that the following types of variations be tried out in the regular classroom:

(1) Variation in learning time, of which examples
are: allowing the slow learner more time on successive
topics and thereby postponing the presentation of some topics
until later in the year or next year;   giving shorter assignments to slow learners;   assigning special homework to slow
learners;   planning arithmetic enrichment for the faster
workers; and allowing faster workers to move to new topics.

(2) Content variations, of which examples are: adding topics for the fast learner that are not ordinarily found
in the course of study such as finding median and mode
while omitting a few topics for the slow learner;   varying
the level of difficulty undertaken on any one topic--for example, encouraging faster learners to master the 10's, 11's,
and 12's in multiplication and division or having the slower
worker do exercises in dividing decimal fractions that involve only whole numbers and tenths as a divisor;   varying
the content of practice exercises;   providing rapid learners
with more difficult horizontal enrichment;   providing for
rapid learners more difficult types of horizontal enrichment
which stimulate an interest in mathematics as a hobby;   and
allowing fast learners to study certain selected topics normally taught in a higher grade--for example, progressing
faster with the learning of addition, subtraction, multiplication, and division facts, or studying the meaning of per
cent in the fourth or fifth grade, or tackling areas of par-
allelograms and circles in the sixth grade.   At the same
time a whole school plan for delaying the teaching of a few
selected topics for the slower learners might be inaugurated
--for example, delaying the counting of groups of 2's, 3's,
and 5's, for a grade or more and delaying multiplication and
division of decimal fractions until after grade six.

(3) Varying teaching methods and materials, of which
examples are: follow-up reteaching of new skill to slower
learners;   frequent review for slow learners;   closely teacher-directed reading of textbook for slow learners;   longer
and frequent use of concrete materials with slow learners;
independent use of textbooks by fast learners;   mental arithmetic exercises for fast learners;   use of the encyclopedia
and other such materials by fast learners to investigate
arithmetic topics;   and differentiated test items for faster
learners.

Types of Class Organization

Each of several types of class organization may offer possibility as an organization in which the teacher can make instructional variations. The following types of class organization have been proposed:

(1) Class-as-a-whole procedure in which the teacher carries all pupils through the arithmetic program for the school year together and gives help and encouragement to individuals as the need and opportunity for doing so are recognized.

(2) Combination of whole class and small group organization. Each new topic is introduced to the class-as-a-whole. The class is later grouped so that the teacher may reteach when necessary, use varied materials, provide practice on different levels of difficulty, and provide different learning activities. When another new topic is undertaken, the class again works as a whole.

(3) Grouping the class in two or more groups according to arithmetic achievement:
   a. One plan might be to keep the class together on the same area of content but teach each group separately and vary the learning activities as needed.
   b. Another plan might be to make an effort to keep the class together as successive topics are studied. At the beginning of the school year, the class is divided into two or more subgroups; each group moves forward at its own rate according to the logical sequence of arithmetic topics.

(4) Completely individualized instruction in which each pupil proceeds from one topic to another at his own rate.

An In-Service Project

A group of 36 teachers attempted to explore the advantages and disadvantages of using certain teaching variations in different types of class organization. In the beginning all of these teachers were using a strict class-as-a-whole plan for teaching arithmetic. They expressed the

feeling that they were giving considerable individual attention
to the slower learners but were not sufficiently challenging
the more able learners.

In order to explore possibilities for making instructional variations within each type of class organization, the
teachers voluntarily agree to use certain types of class organization and to keep a brief record of daily plans.   Seven
teachers agreed to use the combination plan of whole class
followed by small groups as needed (Type 2 above).   Eight
teachers decided to try out a grouping-by-achievement plan
in which the teacher would alternate her time between two
groups.   The middle-average to above-average pupils in
achievement were placed in one group and the low-average
to below-average pupils in achievement were placed in another
group.   Three of these eight teachers decided to let each
group move from one topic to another at its own rate of
speed rather than trying to keep the whole class working on
the same general topic (Type 3b).   Five of the eight teachers
who were grouping by achievement preferred to keep both
groups on the same general topic though each group would
be taught separately while using some variation in methods
and materials and variation in practice as well as enrichment activities (Type 3a).   The remainder, 21 of the 36
teachers, preferred to try to meet the needs of children in
a whole class organization.

Through classroom visitation, conferences with teachers, and reading of written records, the following observations were made:

(1)   Teachers using a grouping plan or a combination
whole class and small group plan seemed to be more alert
to the special needs of both the slow and the fast learners.
As one teacher said, "I discover children's needs better
now. "

(2)   Teachers using a strict class-as-a-whole plan
appeared to be generally more reluctant to try out variations
in content or learning activities.

(3)   Just the idea of whether to group seemed to be
the major question in the beginning.   Later more of the
teachers saw class organization as a procedure which might
enable them to make variations to meet the varying needs of
children.   They came to realize that just the act of teaching
by groups is no answer to the problem of meeting individual

differences.

(4)   These teachers did not feel a necessity to use the same plan of class organization all the time.   Several teachers using a certain plan in the beginning began using two plans for class organization.   The plan of organization varied with the topic or stage of development of the topic. Some teachers changed to an entirely different plan.

(5)   Of the 21 teachers who were using the whole class plan of organization, four expressed the belief that they were not meeting individual needs as well as they might through some work with groups.   At times during the year whenever a need was recognized, these teachers used a two-group plan based on achievement on a particular topic.

The other 17 teachers expressed the belief that they were doing as well in meeting individual differences in arithmetic as could be done with other types of class organization.

(6)   Generally, all of the teachers using some type of class organization which varied from strict class-as-a-whole procedure made occasional time variations and variations in methods and materials.   Content variations were made infrequently.   Evidence of some kind of content variation was more often observed in classes which were not using a strict class-as-a-whole procedure.

(7)   In general, these teachers did not seem to use to full advantage the extensions suggested in the teacher's manual for the fast, average, and slow learners.   They frequently moved on to the next page or topic in the textbook without further extension of the previous topic.   A plan for varying arithmetic activities and content for children on different levels of ability is a new effort for many teachers.

Conclusions

Regarding the problem of meeting individual differences in arithmetic, the following conclusions are made:

Teachers seem to agree that the task of meeting individual differences in the regular classroom makes necessary teaching variations.   Variations in learning time, content and teaching methods and materials seem to offer possibility.

There appears to be a close relationship between the teacher's willingness to vary from a strict class-as-a-whole situation and the use of varied content and procedures which might aid in meeting individual differences.   Less variation in an attempt to provide the extra help needed by the slow learners and the extra challenge needed by the fast learners has been observed in classrooms using a strict class-as-a-whole teaching organization.   Opportunities for meeting individual differences in a class-as-a-whole situation might well receive some careful study since the majority of the teachers appear to teach arithmetic in a class-as-a-whole organization.

Presently a flexible plan for the use of some whole class teaching in arithmetic and some teaching by small groups seems to be favored by the teachers desiring to vary the class organization in an effort to facilitate the use of variations for meeting individual differences.   Teachers willing to try out varied class organization learn to recognize situations or activities in which the class as a whole can work well together and other times when the topic or activity seems best carried out through the use of small groups.

This exploratory investigation of the problem suggests that those teachers who are willing to try out some variations or differentiations are making some progress with this problem of meeting individual differences in arithmetic.

AN ANALYSIS OF INDIVIDUAL DIFFERENCES IN ARITHMETIC

Oscar T. Jarvis

While this investigation was not an all inclusive one, it should assist the teacher in formulating some concepts about analyses of individual differences among his students of arithmetic.   Some of the findings which the investigation did establish that should be helpful are listed accordingly.

Range of individual differences at the sixth-grade level:

(1)   The overall range of differences in arithmetic achievement among sixth-grade pupils may vary as much as seven years.

(2)   Among the bright pupils with IQs of 115 or more the range of arithmetic achievement is about four years.

(3)   Pupils of average intelligence, that is those possessing IQs of 95 to 114, will vary in arithmetic achievement by about five years.

(4)   The slow children with IQs of 94 or less may vary as much as five to seven years in achievement.

Percentage of pupils working above, at, and below sixth-grade level:

(1)   When considering the arithmetic achievement levels of all types of pupils--bright, average, and slow--one may expect to find about 69 per cent of them working above, 11 per cent at, and 20 per cent below grade level.

(2)   Among the bright students about 94 per cent will be working above, 4 per cent at, and 2 per cent below grade level.

(3)   About 74 per cent of the average students will be working above, 13 per cent at, and 13 per cent below grade level.

(4)   Approximately 37 per cent of the slow pupils will be working above, 14 per cent at, and 49 per cent below grade level.

The fact that there exists in the elementary school a wide range of individual differences in the area of arithmetic is an incontestable fact.   The teacher cannot and should not seek to eliminate them.   But what he does to meet these individual needs once they have been identified is the important issue and should be the goal to which good arithmetic teaching is directed.

XI. HISTORY AND BACKGROUND

The additional information contained in the excerpts
that are included in this final chapter should be particularly
pertinent to the reader who has first read the chapters preceding it.

E. Glenadine Gibb

The first passage in the last chapter of this volume
is excerpted from an article written for the May 1970 Arithmetic Teacher by Professor Gibb of the University of Texas,
where she teaches mathematics education.

As a former editor of the Arithmetic Teacher Professor Gibb is well known to readers of literature concerning the teaching of mathematics at the elementary school level.   The scholarship shown in this article is a mark of all
her writing.   This article emphasizes once again that there
are many approaches that have been tried by those wishing
to use the process of individualization in their teaching.
Some of these approaches have been well publicized and
some require an elaborate structure but, as is pointed out
in this article, a great variety of approaches have been developed and used in single classrooms, single schools, and
single districts.   Information about these approaches has
not been disseminated widely (or at all).

The questions asked by the author in the last part of
her article seem to me to be particularly relevant.   I should
consider my selection of material for this volume to be successful to the extent that the reader recognizes that these
questions have previously been raised either explicitly or
implicitly.   My attempt has not been to find material that
includes specific answers to all these questions, partially
because pat answers of universal application are not necessarily to be found and especially because it is my belief that

books of this kind are more useful when they start the readers' thinking processes than when they attempt to furnish answers to every question.   No one should be asked to accept
answers merely because they have been given by so-called
authorities.   Individualized instruction not only sets the children free but also unshackles teachers, curriculum makers,
and supervisors of instruction so that they can be as creative
as they are able to be.

Paul Douglass

        The second article is written by one whose experience
is so broad as to almost compel attention to his ideas.   Professor Douglass, who presently is professor of government
and director of the Center of Practical Politics at Rollins
University in Florida, spent 12 years as the president of
American University in Washington, D. C. , has written extensively on variety of subjects, and has held innumerable
positions calling for a high degree of responsibility.

        Professor Douglass' article presents a strong argu-
ment for individualized education without glossing over the
difficulties inherent in changing curricular procedures.   He
also warns of the uselessness of embarking on a course of
individualization merely to be in step, without any effort to
make a real change from conventional procedures.   The ten
principles he draws from the studies of William S. Learned,
whose biographer he is, are worthy of substantial study and
consideration.   They constitute an argument in favor of individualization that seems to be very hard to answer.

        When Douglass speaks of evaluating an individualized
instructional program he emphasizes the need for a willingness to plan for and nourish such an approach if it is to be
successful.   He insists that there must be a firm commitment to individualization if it is to be usefully implemented.
The institution must through its faculty and administration
furnish leadership and guidance to the participating students,
else little benefit will be realized.   Lastly, the various difficulties posed by the individualized approach are succinctly
and clearly stated.

        Readers should not consider lightly the thinking behind
this article because its principles, I am convinced, are applicable to individualization at every level of schooling.   It
would be a mistake to be diverted by the terminology used,

which is directed solely to the college level.

Raymond J. O'Toole

A particularly appropriate follow-up to the passage by Professor Douglass is the very brief excerpt from a 1966 Colorado State College doctoral research study by Professor O'Toole, who is now with the Department of Elementary Education at the University of Arizona at Tucson.

It is obvious from the author's conclusions that his effort to individualize instruction was unsuccessful, in the situation described by him, in showing that 14 weeks of attempted individualization would yield more favorable results than a conventional approach. There is nothing about such a study which is repugnant to the believer in individualized instruction. No experienced or intelligent reader would expect complete success to result from every single attempt to individualize instruction. The explanation of this result is neither here nor there but it certainly can be hypothesized that some of the "perils" cited by Professor Douglass in the preceding article may not have been given full consideration.

Robert E. Botts

Equally appropriate as a follow-up to the Douglass article is an excerpt from an article written for the Journal of Secondary Education by one with broad experience in the field of education. The article itself is so straightforward and so logically written that it is satisfying to know that it was written by an active educational practitioner. Mr. Botts secured his B. A. and M. A. degrees from the Universities of Missouri and Southern California. After a period of teaching and counseling at the junior and senior high school levels, Mr. Botts turned to the area of administration, in which he has been involved for the past 13 years. Presently, he is principal of the Boyd and Reid High Schools in Long Beach, California as well as coordinator of the guidance (opportunity) classes at those schools.

Mr. Boyd has been active in teaching workshops and serving as a consultant about individualizing the approach to learning; he has written a short book about this approach with Donald R. Reid, entitled Individualized Instruction, which

deals with "how to write behavioral objectives, " "how to outline a course using behavioral objectives, " and "how to outline a course using behavioral objectives and instructional
packages. " I have read this material and found it both
practicable and informative.

It certainly must be clear at this point that the article in question is not a theoretical discourse by one who is
unfamiliar with the day-to-day "real life" in the school.   As
I see it this article makes six important points which I commend to the reader as worthy of careful consideration:

(1)   Students must play a role in setting goals and
these goals should serve as referents to the evaluator of the
individual's learning activity.

(2)   There must be available a variety of materials
for use by students and the use of a variety of media by
both students and teachers should be encouraged.

(3)   Education should not be about yesterday or tomorrow but should be relevant to today.

(4)   Creativity by students should be encouraged but
the surest way to bring it about is by giving the student a
sense of accomplishment.

(5)   A sense of accomplishment is fostered when
goals or behavioral objectives are clearly defined to both
student and teacher.

(6)   The successful teacher using the individualized
approach must feel secure in the use of a variety of instructional techniques and methods.   It is the function of such a
teacher to establish a positive climate leading to greater
creativity on the part of all concerned.

Lucille Lindberg and Mary W. Moffitt

I am intensely pleased to conclude this volume by re-
printing an article written by two professors of education at
Queens College who are my highly respected and admired
colleagues on the staff of the City University of New York,
of which both Queens College and Brooklyn are a part. Their
article says in a very effective way exactly what I have had
in mind in the compilation of this book, so that I will not
belabor my readers with any further comment.

THROUGH THE YEARS:
INDIVIDUALIZING INSTRUCTION IN MATHEMATICS

E. Glenadine Gibb

Let us assume that individualized instruction provides ways to teach a group of students so that each pupil can take what is for him the "next step" in his development of mathematical understandings and competencies at the time when he is ready to move ahead. Individualizing instruction requires developing ways to permit the student to progress at his own rate according to his own style of learning and ways to motivate him to think creatively in formulating his mathematical concepts and knowledge of mathematics.

In the field of elementary education, selected headlines relevant to the concern for individualized instruction through the years capture one's eye. Among these are the ungraded school, promotion plans, the Burk plan, the Winnetka plan, the Dalton plan, the contract method, ability grouping, departmentalization, the nongraded school, the Program for Learning in Accordance with Needs (PLAN), Individually Prescribed Instruction (IPI), Comprehensive School Mathematics Program (CSMP), continuous progress, and computer assisted instruction (CAI). Also, there have been local efforts not so widely publicized whereby schools, individual teachers within schools, and school systems have attempted to resolve the problem of individualizing instruction in operational terms for children in their schools and classes. Can lasting breakthroughs be made in an effort to resolve this ever-persistent problem? What will lie in the mysteries of the future for individualizing instruction in mathematics?

During the colonial and early American period, schools were ungraded and most instruction was tutorial in design. This enabled each child to progress at his own rate through those few texts that were available. By 1870, however, with pressures to educate more children, nearly all elementary schools in the United States were changed from ungraded to graded systems. This movement was conceived and established in the faith that all men are created equal. Graduates of normal schools, although lacking an understanding of child development and individual differences, were confident

of what was to be done and what was to be learned at each
level in the graded school.

Throughout the years, areas of "certainty" in mathematics have been maintained.   Guided by textbooks and by
teacher-education programs in the colleges and universities,
teachers have attached knowledges, skills, attitudes, and
abilities to each grade level.   What does one do with children who have already obtained knowledge and skill before
they were supposed to do so?   What does one do with children who have passed through a lock-step graded system
and lock-step mathematics program without acquiring the
certain required skills or abilities they were supposed to
have?

By 1875, numerous means of promotion were used
for adjusting the grade placement of children in order to
accommodate learning differences.   If children had not
achieved the "standards" of a grade level, they were not
permitted to move ahead to the higher grade.   If they were
too far advanced, they were permitted to skip a grade.   And,
by increasing the amount of instruction for slow-learning pupils by employing extra teachers, out-of-school tutoring, and
summer classes, it was possible to get more children through
the required curriculum.

The current individualized instruction movement began
in 1888 in Pueblo, Colorado, when Preston Search not only
advocated but practiced a program of individual instruction.
When the first educational tests were given in Detroit schools
in 1910, Stuart A. Courtis noted that the data secured made
clear both the inefficiency of mass methods and the need for
adjusting work to meet individual needs.   He stated, "The
only conclusion to be drawn from these results seems to be
that improvement in arithmetic must be brought about through
some device that will reach each individual and enable him
to progress at his own rate. "   Courtis assumed that
no progress is made if one does not master each phase as
measured in some way.   Practice tests, designed to enable
students to do their own correcting, were constructed.   Thus
each individual was permitted to progress at his own rate.
The teacher received papers only from those students who
could not find their own mistakes.   Furthermore, the self-
scoring devices, daily individual records, and graphs that
were used, served to motivate children to be in charge of
their own development.

In 1912 Frederick Burk (San Francisco Normal School) initiated so-called self-instructional bulletins in an effort to enable every student to progress as rapidly as his individual ability permitted. The arithmetic curriculum was divided into short-step "goals," each goal representing one specific principle to be mastered. Carefully graded explanations of new steps were written in simple language so that a child could proceed individually with as little or as much guidance as he needed from his teachers. Again, self-corrective tests were used to reveal to the student his weaknesses or strengths on any unit of work. Special supplementary exercises were available for drill on each specific difficulty. When all tests were passed, the student received a promotion slip. The amount of time needed to complete a grade of work varied with individuals since each student progressed at his own rate, some taking longer to complete certain goals than others.

One of Burk's faculty members, Carleton Washburne, moved on to accept the position as superintendent of the Winnetka (Illinois) schools. Although the Burk plan had worked in a laboratory school setting, could it be effective in a larger school system? Success of the Winnetka plan was attributed to "whole-hearted, clear-headed, and cooperative efforts" of carefully selected teachers. Teachers spent time teaching, helping individuals or groups, encouraging and supervising. They no longer sat at their desks but were among their students as they worked. No child ever failed nor did he skip a grade. The student began in September and worked at this individual pace until school stopped in June. The school program was divided into knowledges and skills that everyone needed to master, and art and shop courses that provided opportunity to develop individual interests and abilities in group efforts.

In developing knowledge and skill, teachers discontinued recitation methods in favor of a system whereby each student prepared his unit of work with an answer sheet. After a group of units was completed, the student used practice tests to test himself. If practice tests were 100 per cent right, the student asked for the real test. If he did not attain 100 per cent on the real test, he returned to practicing until he felt ready to try the real test again. Upon mastery of a goal, the student worked toward the next goal.

As reported by Washburne, the Winnetka plan

saved time and allowed for a broader and deeper education.
Individual promotions appeared to decrease retardation and
"overage-ness, " and to increase efficiency in tool subjects.
There was no evidence that it cost more to individualize instruction using the Winnetka plan.

Other efforts to break the lockstep graded system
were also being made during these years.   Among these was
the Dalton plan under the direction of Helen Parkhurst (Massachusetts).   The Dalton plan involved freedom, cooperation
and interaction of groups, and learning to budget class time.
The curriculum was individually paced and emphasis on personalized contracts and self-corrective practice materials.
The work for each grade in each of the academic subjects,
beginning with the fourth grade, was laid out in a series of
related jobs or contracts.   Each job was to be done within
a school month of 20 days.   The contract was completed
across all subject areas before a student progressed to any
subject area in the next job.

In England, Jessie Mackinder was individualizing work
for those children entering school.   Using concrete materials,
the working of a new process was shown to groups of children.   Children were then left to work many examples until,
having grasped the underlying principle, they discarded the
apparatus in their own time.

During these years the one-teacher, eight-graded
rural schools were not forgotten.   Brown (Connecticut) and
Hoffman (Illinois) realized that the objective of individualized
instruction was to teach the study of arithmetic and not to
hear recitations.   These leaders believed that the lack of
specially-constructed textbooks for individualization was not
a stumbling block.   Textbooks could meet the conditions if
reasonable care was taken in the selection of these books
to be used.   Each pupil kept his own assignment book and
progress sheets in loose-leaf covers.   If a grade of B was
made, the teacher gave supplementary assignments.   At the
end of each day, each student gave to his teacher the written work completed.   Keeping record sheets and assignment
books was found to be a powerful motive for good work.
So individualizing instruction reached its peak in the latter
1920s.

With the exception of the Winnetka plan, plans for individualizing instruction seemed to pass out of existence in
the 1930s.   This change was due primarily to a shift of

emphasis in the elementary school. Major concern in the preceding years had been on mastery of subject matter, the child now became the focus of attention. Led by Dewey and others who were concerned with an education closer to child life than that presented in the then current subject-centered curriculum, the attention of leaders turned away from subjects. Although the "progressives, " as most of the innovators of that time were called, emphasized the importance of attention on the individual child, little in the way of specific instructional materials designed to promote individualization of instruction was produced.

Following World War II, the focus was again on subject matter in the elementary school curriculum. Emphasis was placed on making mathematics more meaningful to children. Also, developing out of the context of military instruction were ideas of programmed instruction, programmed learning, automated instruction, programmed materials and teaching machines. The idea of programming and systems development became popularized as a process of determining empirically a sequence of interactions or operations to assure a dependable performance at an established standard. Programmed instruction became another means of providing for individualized instruction in the 1950s.

Among those who accepted the challenge of guidance in the improvement of mathematics programs for elementary schools in the 1950s and 1960s were those persons identified with such innovative mathematics programs as: the Stanford Project, the University of Illinois Arithmetic Project, the Madison Project, the School Mathematics Study Group, the Greater Cleveland Project, the Entebee Project, and the Nuffield Project. Enrichment activities, selected problems, variations in computational techniques, suggestions in teaching guides, supplementary materials, programmed materials, and mathematics laboratories have been and are being used to provide at least minimal resources that can assist the teacher in providing for individual differences in mathematical ability and interest.

Improved mathematics programs with new intent on content have been accompanied by explorations in varying the organizational design of the school and its curriculum. These include the nongraded school, increased provision for independent study, team teaching, and adaptations of departmentalization. In addition, advancement in the field of technology has made available many new teaching aids such as

computers and educational television.   New designs in school
buildings have made "innovation" a more attainable goal.
Mass methods still prevail.   Teachers still stand before one
group of children explaining the page in the book, often without understanding themselves the specific learning objective
for which it was designed.   Completing the textbook, page
by page, regardless of how it is done, is still used as the
indication that children have indeed learned.   Despite such
practices, lip service continues to flow freely, proclaiming:
"education is a personal, individual process"; "the individual, not the group, learns"; "the purpose of education is to
develop the individual. "

In the 1960s, the renaissance of individualized instruction brought to the forefront: the need to define objectives and to state them in behavioral or performance
terms; the need to develop both premeasurement and post-
measurement and assessment devices for monitoring progress in the attainment of each objective; and the need for
procedures for planning each individual's mathematical program in terms of the learning objectives of mathematics
programs.   Projects such as Comprehensive School Mathematics Program (CSMP), a Program for Learning in Accordance with Needs (PLAN), Individually Prescribed Instruction (IPI), computer-assisted instruction (CAI), programmed
learning and local school projects have addressed themselves
to revitalizing earlier attempts to individualize instruction in
mathematics and to identifying the objectives of this type of
instruction.   The instructional model of IPI, for example,
makes use of placement tests to locate the individual child
in the continuum of prescribed learning objectives in terms
of observable student behavior.

If education in mathematics is to be truly individualized, then what type of instructional materials make for
most progress in attaining the goal of individualized instruction for all children?   The answer to that question lies in
the future.   It would seem, however, that survival or non-
survival is dependent upon the success or failure to resolve
other problems that have confronted teachers of mathematics
throughout the years.   Among such problems are:

(1)   Can agreement be reached as to what mathematical understandings, skills and competencies are required in
mathematics?   If so, can these goals be stated in terms of
learning behaviors in such a way as to avoid substituting
memorization for indepth learning?

(2) How can one provide the stimulation of thought and cognitive growth in mathematics of each student by asking the "right" questions at the "right" time?

(3) Can research be designed so as to contribute knowledge concerning how children develop mathematical concepts and skills? Certainly, more knowledge is needed about ways to identify learning styles, aptitudes, and interests of individual children, and about how to identify components of an individual learning style.

(4) What instructional strategies can and should be employed in order to predict with confidence that children will develop the abilities to think independently, to make choices, to plan, and to evaluate? Can instructional strategies be designed to match with individual learning styles and learning potential? Can the curriculum be adapted to the needs of students, instead of students being adapted to fit the curriculum?

(5) Is it possible to design suitable materials and at the same time not make the "package" so expensive that their use is prohibitive? Through the years the lack of appropriate materials has been noted as a handicap in moving toward individualized instruction programs.

(6) Can the instructional programs of our schools provide guidance for the administrative organization and structure of our school buildings rather than be restricted because of them?

(7) What changes must be made in teacher-education programs to truly prepare prospective and experienced teachers to assume their responsibilities in the career they have chosen? Can plans be implemented so that the aptitudes, interests, capabilities, and learning styles of adults can be used to maximize each individual's potential as a teacher of children? Repeatedly, deficiencies in the teacher's education to teach mathematics have been highlighted as a barrier in implementing improved programs in mathematics. Obviously, efforts to effectively individualize instruction are dependent upon the teacher, who likely has never experienced professional preparation in the individualization of instruction. Just as materials, planning, organization, and opportunity are needed in the schools, these same needs exist in teacher-education programs. Each teacher must be prepared to accept his responsibility both

as an individual and as a member of a group.

Each of these areas of general concern contains many more specific questions for which answers must be sought. Research in mathematics education has made little headway in these areas in the past. Will more progress be made in the future? One can speculate that educators in our schools, colleges, and universities will continue to strive for educational ideas and engage themselves in those scholarly efforts to help both children and teachers realize their full capacities. The realization of those goals lies in the future.

THEORY, PRACTICE, AND PERILS
OF INDEPENDENT EDUCATION

Paul Douglass

Mounting population pressures and staggering predictions of enrollments in the next revolutionary college decade turn apprehensive academic administrative eyes toward the simplified paradise envisioned in curricular reform. Is there perhaps not a cheap-and-easy way out known by its generic patent nostrum name of "independent study?" Independent study itself bears descriptive sub-titles such as "more learning for less teaching, " "lower costs through self-education, " and "placing the educational responsibility squarely on the student. " With students on the long side of supply and teachers on the short side, simple administrative economics demand, so we are told, optimum employment of the scarcer resource with a mystical tribute to the end product, the competent, informed, responsibile citizen.

A historic study of the outcome of reform by curriculum supports the statement, "tinkering with the curriculum is <u>not</u> the answer in the 1975 problem either in theory or in practice. " The gibberish incantations of the sorcerers now stirring the academic brew need to be studied in the clear light of the theory of individualized education.

The Theory

Although we have lost the main highway, a general

theory of individualized education does exist. For the first
half of our century this theory was developed by the most
extensive and painstaking educational research ever performed. As a staff officer of the Carnegie Foundation for
the Advancement of Teaching, William S. Learned pioneered
an educational theory which finally led, to mention one result, to the establishment of the Graduate Record Examination and Educational Testing Service at Princeton. Now
rich, fat, and magnificent on its Princeton campus, ETS
lives on in a generation "which knows not Joseph. "

What are the fundamental educational propositions in
individualized education when defined in the terms of the
mainstreams of Learned's studies?

True intellectual goals should displace time-serving
formulas.

The mainstream of the students' learning should move
in its own broadening channel, independent of courses, teachers, and institutions.

Liberal education is a satisfaction of a sound feeling
of the relative values that permanently concern human life
wherever they may be found.

A curriculum is nothing more or less than the actual
sequence of mental acquisitions of a given mind as it makes
its way through the world of ideas by which it is surrounded.

Since every mind behaves in a different manner,
every curriculum is necessarily different for every student.

A curriculum presented to the student as a series of
courses is ineffective.

Education is a continuous intellectual voyage to be
given direction by measurement of present worth.

The student experiences intellectual momentum when
with the acceptance of responsibility for self education he
marshals all the emotional and moral qualities of his nature
behind his intellectual task.

No educational design can elevate intellectual performance above the level of the flow from a student's mind.

The library is the active intelligence center on the
campus.

The corollary of the indisputable fact of individual differences is individualized education. Think of the doctor
studying the daily clinical record before he engages in his
daily task of evaluation, diagnosis, and prescription. Think
of the farmer with his dairy herd improvement records concerned with the care and feeding of cattle as productive

individuals. It is a sad commentary on our age that we
have more adequate technologies and their implementation
for the care and feeding of cows than we do for the concern
and guidance of young minds.

The Practice

We can reduce the evaluation of a program of individualized education to a few behavioral questions. For example:

(1) Does the administration of this institution have
an implemented concept of individualized education?
(2) Are faculty meetings and their agenda developed
to center concern on the individual student? Is the primary
educational environment concerned with the motivation of the
student?
(3) Is the information fed back into the educational
process?
(4) Does the institution have available a resource
bank of technologies to support the teacher in his personal
association with the student and help him in his continuous
formulation of his intellectual purpose in terms that he can
understand and that will fire his imagination?
(5) How is the academic environment designed to
communicate to the student that his real business is systematic and profitable thinking?

(6) Does the administration generate constructive
situations in which a student can demonstrate and can do?
Is the budget designed to be supportive of such a program?
Does the faculty join regularly in formulating a carefully
confirmed and worded estimate in educational terms of the
demonstrated abilities of each student: where they lie and
of what nature they are?
(7) Has the institution developed a careful statement
about the fields of knowledge so that a student can use it as
a map to see the terrain and destination--where he is going
and how he gets there?
(8) Are these maps supported by annotated reading
lists?
(9) Do they emphasize trunkline ideas and junction
points?
(10) Finally, does the institution help the student plan
his continuous intellectual voyage and regularly measure his
present worth? Does the institution have an on-going

interest in the performance of the student as he continues up
the educational ladder and into his life career?

The Perils

Monitored education is a primrose path. It is planned,
clocked, designed in Carnegie units, and provides interchange-
able parts for mass production. It is almost fool-proof and
so ingenious as to outwit the most cunning and devious student. Individual education is otherwise. It is neither neat
nor uniform. It is, moreover, subject to perils of its own.

The first of these lies in the student's capability of
organizing and managing his own time. He needs help to
budget his day and week. If he sleeps later, misses meals,
and goes to bed later, he may develop a student folkway
which tugs him down into the academic jungle of academic
underachievement. The second peril of independent study
lies in inadequate mapping of the student's course. He
needs to have his own travel plan, his own itinerary in the
world of ideas. We do not get to Rome by TWA without
known points of departure, airlines, as well as destination.

The third peril of independent study lies in the homogeneous assumption that all students possess instant readiness in goals and motivation as well as in technical capabilities. The fourth peril lies in the adequacy of budgeted
and available resources. Have reference librarians been
added? Is field work properly planned and guided? The
fifth peril lies in the inadequacy of instruments to measure
the student's performance.

The sixth peril lies in the infrequence of student-fac-
ulty contact. We do not stimulate and inspire students by
widening academic distance. If we do, then no reason
exists for the residence college. And finally, we need
frankly to recognize that some students prefer the primrose
path of monitored education. Individualized education is not
for everybody. Already it is developing its own canny sub-
culture!

Obsolescence of the Age of Blueprints

Marshall McLuhan points out that an electric environment of instant circuitry has succeeded the old world of the

wheel and nuts and bolts. It has also superseded the era of academic master blueprints. As W. B. Years expressed it: "The visible world is no longer a reality and the unseen world is no longer a dream. " In an age of fluidity fixed curriculums rest their case on units of classified informa- tion in which the student can find little point for self-in- volvement. We have, as McLuhan says, entered the era of self-employment and artistic autonomy. Continued in their present pattern, he says, our curricula will "insure a citi- zenry unable to understand the cybernated world in which they live. "

What does this new language mean? Merely this: that the electric environment gives urgency to the practice of individualized education. The theory of individualized education stands ready for utilization at every educational point where teachers open the educational process to the student in the direction of his present worth and counseled self-goals.

LEARNING PROBLEM-SOLVING ABILITIES
THROUGH INDIVIDUALIZED INSTRUCTION

Raymond J. O'Toole

The population for this study, designed to compare two methods of teaching science, consisted of 81 fifth-grade pupils. The pupils were divided into three classes of 27 each. The experimenter taught two of the classes, the "teacher-centered active control" one, and the "individualized experimental" group. A classroom teacher taught the "teach- er-centered passive control" group. The total experimental period, including testing, was 14 weeks.

The science content used in this study by the active control and the experimental groups was modified from the AAAS Science--A Process Approach and the textbook, To- day's Basic Science.   The passive control group's instruc- tion was based on the textbook, Today's Basic Science.   The experimental and active control groups received 150 minutes of instruction each week.   The passive control received 200 minutes of instruction each week.

Each group was pre-tested and post-tested with the following tests: the Stanford Science Achievement Test; the Sequential Test of Educational Progress: Science; the Science Research Associates Interest test "What I Like to Do"; a researcher-constructed problem-solving test; and a researcher-constructed self-concept checklist.

The findings in this study led to the following conclusions:

(1) The individualized method of instruction did not prove superior to the teacher-centered approach in the achievement of science content, overall problem-solving ability, increased science interest, or the attainment of a more positive self concept.

(2) The science programs of this study, stressing the achievement of problem-solving abilities as their prime objective, whether individualized or teacher-centered, were significantly more effective than a teacher-centered program not stressing these same objectives.

(3) The pupils in this study achieved the ability to identify hypotheses and problems equally well, regardless of whether an individualized or teacher-centered approach was used, when the prime objective was the achievement of problem-solving skills.

(4) The teacher-centered science program, stressing the achievement of problem-solving abilities as its major objective, was significantly more effective in achieving the ability to identify valid conclusions than the individualized science program stressing these same objectives, or the teacher-centered program not stressing these objectives.

THE CLIMATE FOR INDIVIDUALIZED
INSTRUCTION IN THE CLASSROOM

Robert E. Botts

Many individualized study programs are upon examination highly structured. They are often an improvement over certain traditional methods, they do not provide a truly individualized program in the sense that the student is personally involved with the teacher in establishing goals for himself and planning a course of action to achieve these goals. Educators have failed in "selling" the curriculum as much as they have failed in making it relevant.

The individualized learning program which fosters a sense of self-esteem and self-worth for its pupils possesses an observable atmosphere of productivity and creativity. Students who are productive for the first time in months or years show a feeling of exhilaration. This feeling, however, is not a continuing phenomenon unless the learning is relevant. The faculty and staff should identify productivity and recognize successful efforts repeatedly. Goals or "behavioral objectives" need to be identified. When the student is involved with, committed to, and intensely aware of a goal, the identification of success is clear and easy.

Creativity must be encouraged and no stigma should be attached to the student who tries without success. Students should be helped by their teachers to choose creative activities which are within their potential. Creativity is involved when one uses his powers of rational thinking, sense of values, knowledge, and feeling to internalize a new idea or concept.

Evaluation must be a continuous process, including frequent communication between the student and teacher, relying heavily upon student testimony both in word and action regarding his interest in the area of learning. The teacher can best appraise the student's interest and his ability through informal classroom processes rather than by formal, pressure-oriented test situations.

A highly individualized learning program requires a rich provision of instructional materials within the school, especially in the classroom. Materials must be available when the individual student reaches the point where he needs them, keeping in mind that within a given classroom there may be a diversity of learning activities proceeding simultaneously. One needs only to visualize such a classroom to see that the quantity of materials must far exceed that which is being used at a given moment.

In summary, much of the positive climate for an individualized instructional program can be created by the faculty and staff. It is a part of a relevant learning situation that the student understand the value of the curriculum and be involved in setting goals for himself. A proclivity for effort exists when the student experiences success. The opportunity to be creative and the encouragement of creativity should be of high priority. A large quantity of instructional material is needed in an individualized instructional program. The evaluation of progress should be in reference to the goals set for and by the student.

WHAT IS INDIVIDUALIZING EDUCATION?

Lucile Lindberg and Mary W. Moffitt

What are we doing to provide for the individual differences of boys and girls with whom we work?

Many books and articles have been written stressing importance of preserving the uniqueness of the individual in a democratic society. Many of us give lip service to the value and need for individualizing education and practice very little of it. Even while we are speaking about it, we continue to work towards uniformity of a set body of content that all children should learn.

Why do practice and theory not come together?

Is part of the difficulty that the term "individualization" is interpreted in so many different ways? In some instances, it is a new word used to justify practices already in operation. There are those who tutor an individual child. For them, this is individualized education. Some teachers pass out special learning materials and let children proceed at their own rates to fill in blanks or look up words. For them, this is individualization. In some classrooms children grouped according to ability recite material assigned to them. This is labeled individualization. There are teachers who feel that they are making provision for individualizing when each child is permitted to select the color of his own paper for a project that all children are making. In some classrooms the teacher helps each child select and evaluate the books he will read. Children construct their own spelling lists and determine their own drills. There are those teachers who plan work of varying degrees of difficulty, when each child may select the task in which he feels he can achieve some degree of success.   These are forms of individualization which are in keeping with that teacher's definition.

Each of us who work with boys and girls needs to examine carefully any suggested approach to individualizing education.   Why is it important?   What do we want to accomplish through it?   When should we do it?

Although there are many ways to individualize teaching, we must be certain that whatever we do is based on the premise that each child is unique.   Not only does he learn at his own rate but he learns in his own special way. Since his background is different from that of any other child, he will approach learning in a manner unique for him.

Self-generating Independence

If we truly individualize educational programs, children will be able to see their responsibilities in learning processes and thus can develop an independence that is dynamic and self-generating.   This independence is achieved when teacher and learner work together in planning and evaluating work to be done.   As Willard Olson has so aptly put it, each child must have an opportunity for seeking, self-selection and pacing.

This does not mean as has often been charged, that in individualization no instruction takes place.   Actually an individualized program calls for more instruction and often more intensive concentration on the part of both the teacher and learner.   Neither should we assume that there is no opportunity for group or total class teaching.   There should be many opportunities for the group to benefit from an exchange of ideas.

In our society we take pride in efficiency, but in an attempt to achieve quality in learning we sometimes resort to an assigning-reciting-correcting process.   We forget that a learner takes into himself that which has special meaning for him.   In fact, since some may never find meaningful material in the classroom, they fail to put forth the effort required to learn assigned tasks.

Premature Exposure

We expose children prematurely to subject matter and to skill developments which they could achieve in less time after they have become a bit more mature. We attempt to get all children to do the same work even though we know that individual pacing is more effective than a uniform rate of coverage for all learners.

One of the procedures for individualization receiving much attention is programmed instruction, which is suggested as a way to provide for packing as well as for self-instruction. But we should ask to what extent, a rote learning strategy is adopted; to what extent the inner relationship and logical structure of ideas are maintained? Is the logic of the program the same for each child? Can all children learn most effectively from the same set of programmed items in the same sequence?

Curiosity and inquisitiveness, qualities of childhood that need to be fostered for creative effort of the individual as well as for achieving the foundation of knowledge, are stifled progressively with routine busy work assignments having significance for someone other than the child. Eventually the qualities needed for creativity become dime through disuse.

Developing Self-concept

Individualization is absolutely essential if we are concerned with developing the self-concept of a child. If his work is always handed to him, he depends on others to make his decisions for him rather than taking responsibility for his own direction. The teacher should see a child as a separate entity, and the child should see himself in this light. Standing out in a group as an individual is quite different from the form of individualization which isolates children to work in solitude on teacher-imposed or machine-imposed tasks.

Many children come away from school feeling that they cannot learn or understand knowledge. They never experience what they can do because they have not had an opportunity to discover for themselves.

Many persons deem it reasonable that, in kindergarten or in extra-curricular clubs, children should have an

opportunity to select what they will study and plan ways of
working on it, but do not feel that the same holds true for
regular classwork.   Whether there will be individualization
of instruction and how much, will depend on the teacher's
perception of what school is.   If a teacher perceives himself as the sole dispenser of knowledge or the one to manip-
ulate the curriculum according to his own logic, it will be
very difficult to let children work in an individualized program.   Many of us who grew up in authoritarian homes and
were not trusted as children find it hard to trust children
and help them develop faith in themselves; find it hard to
recognize that each child will learn more if he has an opportunity to make mistakes in a way that reinforces his
identity as a learner or has an opportunity to discover creative ways to use what he learns.

Do we truly want to develop children who are able to
identify problems relating to their own living; locate and
evaluate resources which can be used in solving them?   If
so, we must learn to tolerate having children in our classrooms engaged in many activities at any one time.

Do we believe that each child is different from all
other children and that each has his own cognitive style?
If so, then we know we are losing time if we attempt to
keep all children on the same track.   Yet, we continue to
do it because we literally do not know how to do otherwise.

Although we cannot do the whole job at once, step by
step we can find ways of dealing with individual differences
consistent with the uniqueness of each child.   It behooves us
to read what others are doing and find out what makes sense
for us.   We are unique too, and each of us must find his
own distinctive way of working.

SELECTED BIBLIOGRAPHY

Included here are full references to all the original
material from which the contents of this book have been excerpted. A few other items which are judged to have special relevance are also included.

Bartel, Elaine V. A Study of the Feasibility of an Individualized Instructional Program in Elementary School Mathematics. Doctoral dissertation, University of Wisconsin,
1965. D. A. 26:5284, 1966.

Bierden, James E. Provisions for Individual Differences in
Seventh Grade Mathematics Based on Grouping and Behavioral Objectives: An Exploratory Study. Doctoral dissertation, University of Michigan, 1968. D. A. 30:196A,
1969.

Botts, Robert E. "The Climate for Individualized Instruction in the Classroom." Journal of Secondary Education
44:309-14, 1969.

Bradley, Richard M. An Experimental Study of Individualized Versus Blanket-Type Homework Assignments in Elementary School Mathematics. Dr. Philadelphia: Temple
U., 1967. D. A. 28:3874A, 1968.

Deep, Donald. The Effect of an Individually Prescribed In-
struction Program in Arithmetic on Pupils at Different
Ability Levels. Dr. Pittsburgh: U. Pittsburgh, 1966.
D. A. 27:2310A, 1967.

Douglass, Paul. "Theory, Practice, and Perils of Independent Education." Improving College and University
Teaching 16:273-76, 1968.

Duker, Sam. Individualized Reading: An Annotated Bibliography. Metuchen, NJ: Scarecrow Press, 1968.

Ebeid, William T. An Experimental Study of the Scheduled Classroom Use of Student Self-Selected Materials in Teaching Junior High School Mathematics. Dr. Ann Arbor: U. MI, 1964. D. A. 25:3427, 1964.

Fischer, Barbara Lee and Louis Fischer. "Toward Individualized Learning. " Elementary School Journal 69:298-303, 1969.

Fisher, Jack R. An Investigation of Three Approaches to the Teaching of Mathematics in the Elementary School. Dr. Pittsburgh: U. Pittsburgh, 1967. D. A. 28:4947A, 1968.

Fisher, Victor L. Jr. The Relative Merits of Selected Aspects of Individualized Instruction in an Elementary School Mathematics Program. Dr. Bloomington: IN U. , 1966. D. A. 27:3366A, 1967.

Flournoy, Frances. "Meeting Individual Differences in Arithmetic. " Arithmetic Teacher 7:80-86, 1960.

Gibb, E. Glenadine. "Through the Years: Individualizing Instruction in Mathematics. " Arithmetic Teacher 17:396-402, 1970.

Gorman, Charles J. "The University of Pittsburgh Model of Teacher Training for the Individualization of Instruction. " Journal of Research and Development in Education 2(3):44-46, 1969.

Graham, William A. "Individualized Teaching of Fifth- and Sixth-Grade Arithmetic. " Arithmetic Teacher 11:233-34, 1964.

Grant, Jettye F. A Longitudinal Program of Individualized Instruction in Grades 4, 5, and 6. Dr. Berkeley: U. CA, 1964. D. A. 25:2882, 1964.

Howes, Virgil M. (Ed.). Individualization of Instruction. New York: Macmillan, 1970.

Howes, Virgil M. (Ed.). Individualizing Instruction in Reading and Social Studies. New York: Macmillan, 1970.

Howes, Virgil M. (Ed.). Individualizing Instruction in Science and Mathematics. New York: Macmillan, 1970.

Jarvis, Oscar T.   "An Analysis of Individual Differences in Arithmetic. "   Arithmetic Teacher 11:471-73, 1964.

Jasik, Marilyn.   "Breaking Barriers by Individualizing. " Childhood Education 45:65-74, 1968.

Jones, R. Stewart and Robert E. Pingry.   "Individual Differences. "   In Instruction in Arithmetic, 25th Yearbook of the National Council of Teachers of Mathemations, Washington, DC: The Council, 1960, p. 121-48.

Kaplan, Abraham.   "Achieving Individualized Instruction. " High Points Winter, 1969, p. 6-8.

Keffer, Eugene R.   "Individualizing Arithmetic Teaching. " Arithmetic Teacher 8:248-50, 1961.

Kramer, William R.   "The Borel I S I Program. "   San Mateo, CA: Borel Junior High School, n. d.   (mimeo), 15p.

Lee, J. Murray.   "Individualized Instruction. "   Education 74:279-83, 1954.

Lindberg, Lucille and Mary W. Moffitt.   "What is Individualized Education?"   In Individualizing Education, Membership Service Bulletin, No. 11-A.   Washington, DC: Association for Childhood Education International, 1964, p. 11-14.

Morris, J. Clair.   A Descriptive Analysis and Evaluation of an Integrated Program of Individualized Instruction in Cedar City High School.   Dr. Provo, UT: Brigham Young U. , 1968.   D. A. 29:2937A, 1969.

Morrison, Maria P.   "Mass Method Versus Individual Method in Teaching Multiplication to Fourth Grade Pupils. " Education 57:345-47, 1937.

Nabors, Cecil T.   The Effect of Individualized Verbal Problem Assignments on the Mathematical Achievement of Fifth Grade Students.   Dr. Houston, TX: U. Houston, 1967.   D. A. 29:1168A, 1968.

Neufeld, K. Allen.   Differences in Personality Characteristics Between Groups Having High and Low Mathematics Achievement Gains under Individualized Instruction.   Dr.

Madison: U. WI, 1967. D. A. 28:4540, 1968.

"New Mathematics Curriculum Called Part of 'Quiet Revolution' in Teaching Methods. " Educational Development 2(2):1-2+, 1970.

Nix, George C. An Experimental Study of Individualized Instruction in General Mathematics. Dr. Auburn, AL: Auburn U., 1969. D. A. 30:3367A, 1970.

Olson, Willard C. Child Development. Boston: Heath, 1943.

O'Toole, Raymond J. A Study to Determine Whether Fifth Grade Children Can Learn Certain Selected Problem Solving Abilities Through Individualized Instruction. Dr. Greeley: CO St. Col., 1966. D. A. 27, 3781A, 1967.

Patterson, J. Marian. "An Observation of Computer Assisted Instruction on Under-Achieving, Culturally Deprived Students. " Journal of Secondary Education 44:187-88, 1969.

Potamkin, Caroline C. "An Experiment in Individualized Arithmetic. " Elementary School Journal 64:155-62, 1963.

Redbird, Helen. "Individualizing Arithmetic Instruction. " Arithmetic Teacher 11:199-200, 1964.

Scanlon, Robert G. Factors Associated with a Program for Encouraging Self-Initiated Activities by Fifth and Sixth Grade Students in Selected Elementary Schools Emphasizing Individualized Instruction. Dr. Pittsburgh: U. Pittsburgh, 1966. D. A. 27:3376A, 1967.

Shaw, Archibald P. "Individualized Instruction. " American School and University 36(1):9, 1963.

Sinks, Thomas A. How Individualized Instruction in Junior High School Science, Mathematics, Language Arts, and Social Studies Affects Student Achievement. Dr. Urbana: U. IL. 1968. D. A. 30:224A, 1969.

Snyder, Henry D. Jr. A Comparative Study of Two Self-Selection-Pacing Approaches to Individualizing Instruction in Junior High School Mathematics. Dr. Ann Arbor: U. MI, 1966. D. A. 28:159A, 1967.

Spaulding, Robert L. "Personalized Education in Southside School. " Elementary School Journal 70:180-89, 1970.

Suydam, Marilyn. Evaluation of Journal-Published Research Reports on Elementary School Mathematics 1900 to 1965. Dr. State College: PA St. U. , 1967. D. A. 28:3387A, 1968.

Thompson, R. B. "Diagnosis and Remedial Instruction in Mathematics. " School Science and Mathematics 4:125-30, 1941.

Weisgerber, Robert E. Developmental Effects in Individual- ized Learning. Itasca, IL: Peacock, 1971.

Whitaker, Walter L. "Individualized Arithmetic - An Idea To Improve the Traditional Arithmetic Program. " Arith- metic Teacher 9:134-37, 1962.

Wolff, Bernard R. An Analysis and Comparison of Individu- alized Instructional Practices in Graded and Non-Graded Elementary Classrooms in Selected Oregon School Dis- tricts. Dr. Eugene: U. OR, 1968. D. A. 29:4397A, 1969.

Yeager, John L. Measures of Learning Rates for Elemen- tary School Students in Mathematics and Reading Under a Program of Individually Prescribed Instruction. Dr. Pittsburgh: U. Pittsburgh, 1966. D. A. 27:2081A, 1967.

INDEX OF NAMES

421